Science's Most Wanted

Selected Titles in Brassey's *Most Wanted* Series

Military History's Most Wanted: The Top 10 Book of Improbable Victories, Unlikely Heroes, and Other Martial Oddities, by M. Evan Brooks

Hollywood's Most Wanted: The Top 10 Book of Lucky Breaks, Prima Donnas, Box Office Bombs, and Other Oddities, by Floyd Conner

The Olympics' Most Wanted: The Top 10 Book of Gold Medal Gaffes, Improbable Triumphs, and Other Oddities, by Floyd Conner

Rock and Roll's Most Wanted: The Top 10 Book of Lame Lyrics, Egregious Egos, and Other Oddities, by Stuart Shea

Science's Most Wanted

The Top 10 Book of Outrageous Innovators, Deadly Disasters, and Shocking Discoveries

Susan Conner and **Linda Kitchen**

Brassey's, Inc.

WASHINGTON, D.C.

Copyright © 2002 by Brassey's, Inc.

Published in the United States by Brassey's, Inc. All rights reserved. No part of this book may be reproduced in any manner whatsoever without written permission from the publisher, except in the case of brief quotations embodied in critical articles and reviews.

Library of Congress Cataloging-in-Publication Data

Conner, Susan, 1950–
  Science's most wanted : the top 10 book of outrageous innovators, deadly disasters, and shocking discoveries / Susan Conner and Linda Kitchen.—1st ed.
     p.   cm.
Includes bibliographical references and index.
  ISBN 1-57488-481-6 (pbk).
  1. Science—Miscellanea. I. Kitchen, Linda, 1977–
II. Title.
Q173.C76 2002
500–dc21

2002008007

Printed in Canada.

Brassey's, Inc.
22841 Quicksilver Drive
Dulles, Virginia 20166

Designed by Pen & Palette Unlimited.

First Edition

10 9 8 7 6 5 4 3 2 1

To the men and women of science

May they continue to dream

Contents

List of Illustrations	xiii
Acknowledgments	xv
Introduction	xvii

GENERAL SCIENCE   1

To Ig-Nobely Go   3
Achievements that should never be reproduced are rewarded.

Unusual Experiments   6
Some experiments contributed little to our knowledge.

A Bit Eccentric   11
No great genius is without some mixture of insanity.

Mad Scientists in the Movies   16
Films have not always been kind to scientists.

Hoaxes and Deceits   19
We've been fooled more than once.

Noted Feuds and Controversies   24
Even scientists can be human.

People Who Gave Their Lives and Limbs to Science   34
Some experiments don't end well.

Noted Women Scientists before Women's Equality   39
Women have risen from data gatherers to Nobel laureates.

Eminent Scientists Who Didn't Win a Nobel Prize   44
Any scientist is eligible; only a few win the prize.

Names That Became Scientific Terms   47
Some scientists' names are used every day.

Creative Couples   51
Partnering leads to good ideas.

Codiscoveries   55
Great minds think alike.

Scientists with Disabilities   61
Physical problems have not stopped some scientists.

Unethical Experiments   65
May these experiments never continue.

Blackballed   70
Those who go against the grain pay the price.

ASTRONOMY   77

Noted Early Astronomers   79
Stellar observers observe in more ways than one.

Ten Planetary Pioneers   84
They looked beyond boundaries to find new planets.

Remarkable Observational Tools   88
From Kepler's eyepiece to the Hubble Space Telescope, we have continued to improve our observational tools.

Early Observational Instruments   93
We have used means other than telescopes to watch the heavens.

Models of the Universe   97
This is a list of concepts, not beauty contestants.

Astronomical Surprises   102
Just when we think we know the answers, someone finds a surprise.

Exoplanets   106
At least 50 planets have been sighted outside our galaxy.

PHYSICS   111

Building Blocks of Matter   113
There's more to learn than just protons, neutrons, and electrons.

Atomic Theories   117
What's discarded is less important, but more amusing, than what's left behind.

The Bomb Squad   121
Members of the Manhattan Project changed the world.

GEOLOGY   127

Ancient Views on How Life Began   129
Are we full of fire or watered down?

Modern Views on Origins   133
Were germs here first?

How and When Earth Began   137
Our world has gone from being viewed as the center of the universe to just another planet.

What on Earth?   141
Theories from a hollow earth to drifting continents abound.

How Old Is Earth?   146
Was Earth created in 4004 B.C.?

Ice Age Theories   **150**
Does Earth's tilt, heavy oxygen, or dust cause climatic catastrophe?

Mineral Monikers   **155**
Mineralogists play the name game.

BIOLOGY   159

Aliens in Our Backyard   **161**
Humans just think they're dominating, but, really, plants control the world.

Invasive Animals   **169**
Some species just don't play fair.

Unusual Reproduction   **175**
Anything goes.

Poisonous Plants   **179**
Berries and mushrooms take their toll.

Unusual Forms of Life   **184**
Not every crevice on the globe is found in the textbooks.

Microlivestock   **188**
The best food on Earth is right under your feet.

Weather-Forecasting Tools   **191**
Mother Nature aids us in predicting the weather.

Worst Extinctions   **193**
Current habitants represent only a fraction of what has been here before.

PALEOBIOLOGY   197

Dinosaur Discoverers   **199**
Scientists often didn't know where to look or what they were looking for, but they scored significant finds.

Recently Discovered Dinosaurs — 206
  Brush up on your dinosaur names.

Why Dinosaurs Became Extinct — 210
  Did dinosaurs go out with a bang or just a whimper?

All in the Family? — 215
  The human tree has added many branches.

CHEMISTRY — 221

Prominent Early Chemists — 223
  Movers, shakers, and big-name players set the stage for the acts ahead.

Alchemy — 229
  The dream of gold gave some scientists nightmares.

Unusual Chemists — 235
  Chemistry is not confined to the laboratory, or to humans.

Living Better through Chemistry — 239
  Students, chemistry is only as difficult as you make it.

MEDICINE — 245

Cancer Suspects — 247
  Researchers line up possible perpetrators.

Cancer Treatments — 252
  Many varieties of cancer require many sorts of cures.

Not-So-Common Cold Cures — 256
  The remedies can be worse than the symptoms.

Fountain of Youth — 259
  Living forever has a high, and sometimes disgusting, price.

TECHNOLOGY 263

Innovations from Outer Space 265
The space program has contributed to our everyday lives.

Artificial Intelligence 269
Soon our pets will make better decisions than we do. Or do they already?

Your Flight Has Been Canceled 273
Our attempts to scale the heights have occasionally met with tragedy.

Computer Advancers 277
All the bit players make up one advanced system.

Internet Accelerators 280
Getting your online in line.

Most Fascinating Systems 283
With time and necessity, humans can create one big mess.

Hacked Off 289
Don't let these computer bandits near your network.

Dot.bombs 294
The other end of a big incline is a big decline.

Bibliography 299

Index 305

About the Authors 316

List of Illustrations

Sigmund Freud 29
(From a drawing by Hermann Struck for *Medicine and the Artist* [Ars Medica] by permission of the Philadelphia Museum of Art.)

M. and Mme. Lavoisier 35
(Courtesy of The Metropolitan Museum of Art, Purchase, Mr. and Mrs. Charles Wrightsman Gift, in honor of Everette Fahy, 1977. [1977.10])

Lord Kelvin 49
(Courtesy of the Mary Evans Picture Library.)

Lise Meitner 71
(Max-Plank Society archives, Berlin, Germany.)

J. Robert Oppenheimer 74
(Courtesy of Emilio Segré Visual Archives, American Institute of Physics.)

Purple Loosestrife 166
(Photograph by Linda Kitchen.)

The *Rat Catcher* 173
(From a drawing by Cornelis Visscher for *Medicine and the Artist* [Ars Medica] by permission of the Philadelphia Museum of Art.)

Sir Richard Owen 201
(From the drawing "Richard Owen, Riding His Hobby," by Fredrick Waddy, *Cartoon Portraits and Biographical Sketches of Men of the Day*, 1874.)

Paracelsus 225
(From a drawing by Balthazar Jenichen for *Medicine and the Artist* [Ars Medica] by permission of the Philadelphia Museum of Art.)

Apothecary Shop 241
(From a drawing by Jost Amman for *Medicine and the Artist* [Ars Medica] by permission of the Philadelphia Museum of Art.)

Acknowledgments

Our sincere thanks go to all who helped us create *Science's Most Wanted*. Floyd Conner, the "Most Wanted" series's originator, encouraged us throughout the project and provided valuable insights. Physicist Marcus Thielen, an international neon consultant, contributed his considerable knowledge of science, particularly physics, while David Jepson contributed his expertise in computer science. Publisher Don McKeon and editor Paul Merzlak set us on the right path concerning the book's tone and organization. Science historian Dr. William J. Astore kept us on course. Also, the helpful librarians and extensive resources of the Hamilton County and Cincinnati Public Library contributed invaluable information resources.

Introduction

In the search for knowledge, humanity has focused its limited observational tools—the senses, as well as microscopes and telescopes—on everything from cells to superstrings. Yet we skim the surface because of the inherent limitations in both the observer and observational tools. We are either too general or too specific, too biased or too inclusive. If an image of the cosmos were beamed to us in its entirety, humans would interpret it in so many ways that we could never settle on one "true" picture. Even if we could, our interpretation would be limited by the scope of our understanding. In other words, we are humans, not gods.

Scientific observations are tested against a theory, hypothesis, or model. Facts never speak for themselves because they are interpreted through ideas and observers. When humans interpreted the world through religious dogma, it was difficult to separate facts from doctrinal resolve, though it must have been comforting to imagine that few things ever changed. After Copernicus, Kepler, and Galileo dethroned Earth as the center of the universe, scientists began to observe the clockwork efficiency of a mechanical universe. When a few glitches—such as the discovery that Mercury's orbit cannot be calculated precisely without incorporating relativistic effects—appeared in the clockwork, we befriended chaos and uncertainty—relativity and quantum theory. Yet there remains the nagging doubt that our observational limitations may play a part in not seeing the complete precision with which the universe works.

Michel Montaigne observed: "Whenever a new discovery is reported to the scientific world, they first say, 'It is probably not true.' Thereafter, when the truth of the new proposition has been demonstrated beyond question, they say, 'Yes, it may be true, but it's not important.' Finally, when sufficient time has elapsed fully to evidence its importance, they say, 'Yes, surely it's important, but it's no longer new.'"

However, science still seems to progress, often providing us with a better quality of life. *Science's Most Wanted* lists not only offenders in the search for Truth, such as hucksters and hoaxes, but also outstanding innovators and heroic skeptics who sometimes faced social and religious ostracism because they dared to see things differently. Assuming an Archimedean perspective of the human condition, in comparing the theories of the ancient Greeks with our modern views of the cosmos, we have found that we possess more physical data than our predecessors. Nevertheless, we still are nearly as far from a complete understanding of "truth" as they were. This book is not written to poke fun at the brave observers of yesteryear but to point out that the history of science is littered with discarded theories.

And this is the only way that science can remain healthy—by correcting biases, slants, and flaws in our interpretations of data. Thankfully, we no longer cling to Aristotle's theory that the world is finite because Earth moves in a circle. Medicine has always reflected the interplay between science and societal issues, but it's far better to be facing the vexing issue of using fetal tissues for research than to stagnate in the belief that we are governed by Hippocratic humors. Antoine-Laurent Lavoisier conducted the same experiment that alchemists used when he heated red oxide of mercury in a glass vessel. But he observed that oxygen was being liberated and collected the gas, noting that chemical decomposition rather than a perfecting principle (the philosopher's stone) was at work. Yet it was Lavoisier who lost his head when the mob of the French Revolution decided to overlook his role as a scientist and condemn him as a mere tax collector—his plea that he was a scientist fell on deaf ears.

Perhaps we don't have the final word on human evolution, but at least we no longer identify Neanderthal fossils as those of a Mongolian Cossack who had fallen while pursuing Napoleon. Robert Plot, professor of "Chymistry" at Oxford, proposed that a weighty specimen found in Oxfordshire was either an elephant brought to Britain by the Romans or a biblical giant, rather than a Megalosaurus.

Certainly, we don't have definitive evidence that shows us how the universe began—we don't even know what it looks like. Some believe that what is here really came from out there. And what is out there may be shaped like a pancake. Or maybe a giant White Wave from a fifth dimension collided with our four-dimensional world to get things moving.

Science has its share of inspiring stories. Dorothy Crowfoot Hodgkin deciphered the atomic structure of penicillin and insulin, despite suffering from rheumatoid arthritis, which crippled her hands and feet. Physicist Edward Teller lost his foot in a streetcar accident but went on to become the father of the hydrogen bomb. The father of modern atomic theory, John Dalton, began life as the son of a poor weaver with a form of color blindness now known as daltonism.

At times, science can be a dangerous pursuit. Italian monk Giordano Bruno was burned at the stake in 1600 in part because he taught throughout Europe that Earth revolved around the Sun. Biologist Nikolai Vavilov promoted classical genetics, opposing T. D.

Lysenko's inheritance of acquired characteristics that supported Russia's revolutionary ideology. For this, he was sentenced to death but spent the rest of his life in prison, where he died of malnutrition.

Scientists—even Nobel Prize winners—possess a sense of humor. The science-humor magazine *The Annals of Improbable Research* awards the Ig Nobel Prizes to such serious research topics as soggy cereal, lilliputian fossils, and drugged clams.

Some animals and plants have won fame greater than those who discovered or experimented with them. They have had amazing and questionable roles in transforming science throughout the ages, while the study of biology has shown the interaction of nature and science. The brandaris mollusk creates a chemical that early Phoenicians harvested to make purple dye, creating the wealthy, yet stinky, city of Tyre. The brown rat invaded and dominated the plague-carrying black rat in Europe, reducing Black Plague outbreaks. And hemlock took Socrates's life when he committed suicide as punishment for corrupting the Athenian youth with new ideas.

This book is not a comprehensive overview of all major scientific theories. We have highlighted a few that represent either breakthroughs in thought or muddled judgment. We hope that we have assembled a varied cast of notables and rogues so that you can nominate your own winners.

GENERAL SCIENCE

To Ig-Nobely Go

Eminent scientists are exalted for groundbreaking work when they receive the Nobel Prize. However, scientists aren't so pleased if they receive the Ig Nobel Prize, presented by the science-humor magazine *The Annals of Improbable Research* (AIR). The award, which honors individuals whose achievements should not be reproduced, is intended to celebrate the unusual and honor the imaginative. Nobel laureates often award the prizes.

1. SILURIAN PARK

In 1996, Chonosuke Okamura of the Okamura Fossil Laboratory in Nagoya, Japan, was awarded the Biodiversity Award for discovering that vertebrate life, including humans, began in the Silurian period. Using slabs of polished limestone from Mount Nagaiwa in the early 1980s, Okamura scrutinized and catalogued minimen and lilliputian animals, which most geologists think are mineral grains and fossils of foraminifera, marine protozoans. Okamura also isolated dragons and monstrous denizens of Mount Nagaiwa.

2. FLAKEY BITS

D. M. R. Georget, R. Parker, and A. C. Smith of the Institute of Food Research in Norwich, England, received the 1995 physics prize for their in-depth analysis of soggy breakfast cereal, published in a report entitled "A Study of the Effects of Water Content on the Compaction Behavior of Breakfast Cereal Flakes." This study was published in *Powder Technology* in November 1994.

3. LOVE CODE

Bijan Pakjad of Beverly Hills, California, scored in the 1995 chemistry awards for creating DNA cologne and DNA perfume, neither of which contained DNA. The scent is decanted in a bottle designed as a double helix.

4. STOP THOSE PIGS!

Biology was revolutionized by the 1970 report, published in the *American Journal of Public Health and the Nation's Health,* by Paul Williams Jr. and Kenneth W. Newell of the Liverpool School of Tropical Medicine that documented salmonella excretion in joy-riding pigs.

5. PAIN MANAGEMENT

James F. Nolan, Thomas J. Stillwell, and John P. Sands Jr. painstakingly reported on "Acute Management of the Zipper-Entrapped Penis," which appeared in the *Journal of Emerging Medicine* in May 1990.

6. STRANGE MOVES

Andre Geim of the University of Nijmegen in the Netherlands and Sir Michael Berry of Bristol University documented their successful levitation of a frog and a sumo wrestler by using magnets. "Of Flying Frogs and Levitrons" was published in the *European Journal of Physics* in 1997.

7. **THE S-CHECK**

Takeshi Makino, president of the Safety Detection Agency in Osaka, Japan, was honored in 1999 for his role in the development of the S-Check, a spray that reportedly detected infidelity when wives applied it to their husbands' underwear.

8. **HAPPY AS A CLAM**

Peter Fong of Gettysburg College in Pennsylvania was happy as a clam to accept the 1999 biology prize for his study of the effects of Prozac on fingernail clams. The *Journal of Experimental Zoology* published the study in 1998.

9. **SPEARMINT OR DELTA?**

T. Yagyu and cohorts from the University Hospital of Zurich, Switzerland; Kansai Medical University in Osaka, Japan; and Neuroscience Technology Research in Prague, Czechoslovakia, received ignoble acclaim in 1997 for measuring brainwave patterns while people chewed different flavors of gum. The full text can be perused in a 1997 issue of *Neuropsychology*.

10. **JUST A PASSING FANCY**

Dr. Mara Sidoli of Washington, D.C., reported in a 1996 edition of the *Journal of Analytical Psychology* that breaking wind was a defense against unspeakable dread.

Unusual Experiments

If scientists didn't go against the grain of conventional thought, there would be no progress. However, the following experiments may not have contributed much to our knowledge of the world.

1. CHICKEN REVENGE

Francis Bacon (1561–1626) died as a result of one of his experiments. Riding in the snow, he was struck by a desire to "try" an experiment or two concerning the conservation and induction of bodies. He purchased a chicken and stuffed it with snow. The chicken remained cold, and the experiment succeeded, but Bacon contracted bronchitis and died.

2. SMOKY CANDLES

French chemist Jean-Baptiste-André Dumas (1800–1884) devised a method for determining nitrogen in organic compounds. At a soiree held by Charles X at the Tulieres, the white candles burned with a smoky flame, emitting irritating fumes. Dumas was asked to investigate the cause of the matter and found that the fumes consisted of hydrogen chloride. The candle factory had used wax bleached with chlorine. A huge argument followed when Dumas stated that hydrogen could be substituted with chlorine or bromine with no alteration of the resulting compound.

3. SHABBY CELESTIAL FURNISHINGS

In 1804, the French Academy of Sciences questioned the Russian report that magnetic force decreases with increasing distance from the earth's surface. Chemist Joseph-Louis Gay-Lussac (1778–1850) and physicist J. B. Brot (1774–1862) volunteered to make an ascent in a hot-air balloon to test the theory. Except for a knife and shears, they carried nothing else but a sheep, rooster, pigeons, snakes, and bees. The balloon achieved only 4,000 meters, but the scientists found variations in the magnetic force. Brot refused to man a second flight, but Guy-Lussac was game. He reached 7,016 meters, a record that stood for more than a century. But the chemist suffered from the cold and had difficulty breathing. During the flight, to attain more altitude, he discarded a white kitchen chair, which fell next to a shepherdess, who screamed for help. The villagers who had gathered initially believed the chair had fallen from the heavenly regions. However, they wondered why such a shabby chair would be used by angels. The mystery was solved when they learned of the balloon flight.

4. BITTER DRAFT

In 1862, French chemist Anselm Payan accused English brewers of adding strychnine to pale ales to increase their bitterness. Sales of the brew consequently plummeted. The brewing company Alsopp & Sons asked chemists A. W. Hoffman and Thomas Graham to advise them on how to disprove the allegations. The chemists showed that if strychnine is added to ale, it can be recovered by soaking charcoal with ale and boiling the charcoal with alcohol. The fact that there were no traces of strychnine in the ale was extensively advertised, thus restoring the ales' previous popularity.

5. **THE FLY WITH THE SWEET TOOTH**

German chemist Robert Wilhelm Bunsen (1811–1899) was in the midst of determining whether beryllium has a sweet taste when a fly alighted on the precipitate and devoured the sweet, sticky substance. When Bunsen caught the fly, which had a huge glob of the precipitate on its proboscis, he killed it, taking care not to touch the substance. Then he cremated the fly and treated the ash with a drop of hydrochloric acid and ammonia water. After evaporation and ignition, the missing precipitate was restored.

6. **CULINARY EXPERIMENTS**

The son of a Prussian general, German organic chemist Adolf von Baeyer (1835–1917) ran an orderly laboratory. In 1875, Baeyer succeeded the renowned Justus von Liebig at Munich University. Liebig had paid no attention to what went on in the laboratory. When Baeyer studied the supply book, he found such items as eggs, flour, butter, and raisins. He asked a longtime employee for an explanation. He replied that Liebig was always devising new infant foods, synthetic coffee, and artificial cream. So the lab attendants followed suit and made artificial Gugelhopf, a rich Bavarian cake.

7. **STEAKED OUT**

A professor of experimental physics at Johns Hopkins University from 1901 to 1938, Robert Wood (1868–1955) was noted for producing diffraction gratings for astrophysical studies. As a student, he used his knowledge of chemistry on the fare of a local diner. Patrons suspected that portions of the steaks that they had ordered but hadn't eaten appeared as beef hash the next morning. One evening, Wood left several pieces on his plate but sprinkled them with lithium chloride, a white compound that could pass for salt. The next morning, he came to the diner armed with an alcohol lamp and a platinum wire. He ordered hash, then pierced some morsels on the wire, lit the lamp, and passed them through the flame. When the meat turned a bright red (from traces of lithium chloride), he confirmed the suspicion that the hash had been made from the previous night's table scraps.

8. DOWN THE DRAIN

During his youth, American chemist Frank Whitmore (1887–1947) studied the organic derivations of arsenic. One experiment produced such a nauseating smell that he quickly dumped it down the drain. After World War I ended, he discovered that the horrible smell had derived from Lewisite (chlorovinyl dichloroarsine), which had been officially discovered by W. Lee Lewis, but too late to be employed as a poisonous gas in the war.

9. GOLD IN THE SEA

German physical chemist Fritz Haber (1868–1934) was awarded the Nobel Prize in 1918 for extracting nitrogen from air to make ammonia as a chemical fertilizer, which provided Germany with wartime chemicals after a blockade cut off the supply of saltpeter. However, his biggest disappointment was his failure to extract large amounts of gold from seawater. Misjudging the concentration of gold in the ocean, he failed in his plans to obtain enough gold to pay Germany's war reparations.

10. MICROVENUS

In 1982, genestheticist Joe Davis walked into the Massachusetts Institute of Technology uninvited. A secretary called the security force to expel him, but 45 minutes later, Davis walked out as a research fellow in visual studies. Around 1986, genetic engineering inspired him to synthesize DNA and insert it into the genomes of

Escherichia coli bacteria. He then set about creating an "infogene," a gene to be translated by humans into meaning and not by the cells into protein. He created the message "Microvenus," a *Y* with an *I* superimposed on it, representing the Germanic rune for life and an outline of the external female genitalia. Digitized and translated into a string of 28 DNA nucleotides, Microvenus first slipped between the genes of *E. coli* in 1987, and the bacteria quickly multiplied. Davis's idea is to encode a sign of human intelligence into the genome of bacteria, which could be transmitted to extraterrestrials. Davis also created the self-assembling clock to see whether the components of machines can self-assemble into working machines.

A Bit Eccentric

"No great genius is without some mixture of insanity," Aristotle stated. The examples below show that there is some truth to the idea that those who are "out of sync" with their environment are driven by a desire to be different in order to prove their worth to society.

1. **WILLIAM HARVEY**

English physician William Harvey (1578–1657) discovered how blood circulates and how the heart functions. His *An Anatomical Treatise on the Movement of the Heart and Blood in Animals* is cited as a foundation for physiology. He worked in the dark because he said he could concentrate better that way, and he constructed underground caves beneath his house in Surrey so that he could obtain the serenity he needed for meditation.

2. **HENRY CAVENDISH**

Physicist and chemist Henry Cavendish (1731–1810) recognized hydrogen as a separate element and discovered that water was a compound, not an element. His many accomplishments and discoveries are often associated with other research. During his life,

Cavendish was stricken with intense shyness. He ordered all female housekeepers to keep out of sight or else be fired on the spot. He also ordered everyone out of the room just before he died so that he would be alone.

3. ROBERT WILHELM BUNSEN

German inorganic and physical chemist Robert Wilhelm Bunsen (1811–1899), like many brilliant people, was a bit forgetful. He loved to attend dinner parties, but he usually forgot the invitations until the next day. He would then arrive at the host's residence at the correct time, although one day late, and act as if everything were normal. The women at Heidelberg learned to go along with the game. They would quickly notify friends, assemble food, and have another party.

4. FRANCIS DALTON

English anthropologist, eugenicist, and statistician Francis Dalton (1822–1911) had an obsessive desire to quantify everything from the curves of women's bodies to the number of brushstrokes used to paint his portrait. He once decided to taste everything in his hospital pharmacy in alphabetical order but stopped at *c*, when he swallowed castor oil and suffered from its laxative effect.

5. OLIVER HEAVISIDE

Nominated for the 1912 Nobel Prize, British inventor Oliver Heaviside (1850–1925) made long-distance phone calls possible with his electric theories. He also predicted the existence of the ionosphere. For the last 17 years of his life, the flawed genius lived upstairs in the home of the unmarried sister of his brother Charles. Oliver kept her a virtual prisoner by not allowing her to leave the premises of her own house without his permission. The man who signed documents as "W.O.R.M." also enjoyed working in tightly shuttered rooms that were swelteringly hot. Not only did Oliver paint his nails cherry pink, but he replaced his furniture with granite blocks.

6. NIKOLA TESLA

Wearing a black Prince Albert coat, derby hat, white silk handkerchief, and stiff collar even in the laboratory, Nikola Tesla (1856–1943) invented alternating-current power transmission and the induction motor. Each time the Serbian-American inventor made a reservation, he asked the maitre d' to stack 18 napkins in a neat pile, and he would then lift each napkin and discard it before dining. Among many other eccentricities were an extreme aversion to the sight of pearls; a hypersensitivity to camphor; and an obsessive love of pigeons, particularly white doves. Even when evicted and penniless at the end of his life, Tesla had hidden away $5 for birdseed.

7. FRITZ ZWICKY

Swiss astronomer and physicist Fritz Zwicky (1898–1974) discovered 18 supernovae and argued that they are completely different from ordinary novae. And argue is what Zwicky did best. He accused his mentor of never having had a good idea. When he taught at the California Institute of Technology in Pasadena, he would say to students whom he didn't know, "Who the hell are you?" According to Zwicky (pronounced "svicki"), other astronomers at Mount Wilson were "spherical bastards" because "they are bastards when looked at from any side." He often arose during other professors' lectures and announced that they knew nothing and that he himself had the answer. When a group of students appeared at his house at his invitation, his wife, Dorothea (whom he divorced), called to him, "Fritz, the bastards are here." During a particularly turbulent night at Mount Wilson, he told an assistant to shoot a gun into the air to smooth things out. It didn't.

8. RICHARD FEYNMAN

For almost two decades, American theoretical physicist Richard Feynman (1918–1988) taught a course called "Physics X," in which students could ask Feynman any scientific question they wished. Working on the Manhattan Project, Feynman developed a theory of predetonation that measured the likelihood that uranium might explode too soon. Fascinated by rhythm and sequence, he played bongos and absentmindedly drummed his fingers and tapped on walls and garbage pails. He also made columns of ants march when the mood struck him.

9. ALAN TURING

English mathematician Alan Turing (1912–1954) played leading roles in breaking the Germans' secret Enigma Code in World War II, which gave the Allied forces advance knowledge of German movements and battle plans, and later in developing the modern computer. He also developed the Turing Test, which measures the "intelligence" of computers. Throughout Turing's life, he was naively straightforward, openly atheistic, and homosexual. In the 1930s, he found refuge in mathematical research and in the intellectual atmosphere of Cambridge. In 1940, he decided to protect his savings, held in silver ingots, against the disaster of another war. He buried the ingots, thinking he could recover them after the war. He buried one in the woods near Shenley and another under a bridge in the bed of a stream, and then enciphered instructions on how to recover them, though he never found them. At Cambridge, Turing wore a gas mask as he bicycled to school to protect himself from pollen because he suffered from hay fever. There were also stories of his trousers being held up by a string and pajamas worn under a sports coat. Never one to attend to personal hygiene, he had a permanent five o'clock shadow and picked at the sides of his fingernails, leaving scars.

In 1952, Turing was arrested for "gross indecency" after reporting a burglary that he had suspected was committed by a lover or the lover's acquaintance. He made the mistake of being blithely informative to the police about his homosexual activity. Afterward he submitted to psychoanalysis and hormonal treatment with estrogen designed to cure his homosexuality. In 1952, he committed suicide by cyanide poisoning.

10. **KARY MULLIS**

The ease with which we reproduce DNA comes from the work of American biochemist Kary Mullis (b. 1944), who invented the polymerase chain reaction technique for analyzing DNA. After winning the Nobel Prize in 1993, he celebrated by going surfing. Describing himself as a "loose cannon on deck," Mullis recounts some of his more bizarre experiments in his autobiography, *Dancing Naked in the Mind Field.* Alone in a deserted cabin in the woods, he lost his sense of time and place while under the spell of a "talking, glowing vacuum." He claims that this episode may have involved alien abduction.

Mad Scientists in the Movies

Cinema has not been kind in its treatment of scientists. Boris Karloff starred in the 1931 movie *Frankenstein,* with actor Colin Clive playing Baron Frankenstein, the mad scientist who brings the monster to life from parts of dead bodies. There have been many screen versions of the Robert Louis Stevenson classic *Dr. Jekyll and Mr. Hyde.* In 1920, John Barrymore amazed audiences by making the transformation from Dr. Jekyll to his evil alter-ego Mr. Hyde without the use of makeup, and in 1931, Frederic March won an Academy Award as best actor for his performance. Based on the H. G. Wells's classic novel *The Island of Dr. Moreau, The Island of Lost Souls* starred Charles Laughton, who eerily portrays the mad doctor whose diabolical surgical experiments transformed humans into a grotesque array of half men, half beasts.

1. ***DR. CYCLOPS***

In this 1940 classic, Dr. Thorkel, an evil scientist, played by Albert Dekker, shrinks people to toy size.

2. ***MESA OF LOST WOMEN***

Jackie Coogan stars as a mad scientist who creates a giant tarantula and a race of superwomen. This 1952 clinker features a score by Ed Wood.

3. ***BRIDE OF THE MONSTER***

Bela Lugosi plays Dr. Vornoff in this 1956 movie directed by Ed Wood. Bela, assisted by the dim-witted hulking Lobo (played by Tor Johnson), tries to create a race of superbeings through the use of atomic energy. A highlight features Bela wrestling a rubber octopus in a mud hole. Subjects of experiments are transformed after being strapped to a table and having a lamp shade placed over their heads.

4. ***THE FLY***

David Hedison plays scientist Andre Delambre, whose matter-transmitter experiments transform him into a half man, half fly. At the end, he is reduced to fly size with a fly's body but his own head and is caught in a spider web.

5. ***THE NUTTY PROFESSOR***

Jerry Lewis's best movie, this 1963 comedy features Lewis as Professor Kelp, a Dr. Jekyll–type klutz who drinks a potion and turns into Buddy Love, a slick, obnoxious ladies' man. Eddie Murphy stars in the 1996 remake.

6. ***DR. NO***

Sean Connery played James Bond for the first time in this 1963 flick. Dr. No (played by Joseph Wiseman) runs a secret island base with a diabolical plan to destroy rockets launched from Cape Canaveral.

7. *THE BEAST OF YUCCA FLATS*

Tor Jonson plays a Russian atomic scientist who walks into an atomic blast. He becomes a monster who walks through the desert, kills people, and eats their hair. Ed Wood directed this 1961 bomb.

8. *DR. STRANGELOVE*

In this 1964 Stanley Kubrick masterpiece, Peter Sellers plays Dr. Strangelove, a German ex-Nazi nuclear physicist who advises the president during a nuclear attack. Strangelove can never quite control his mechanical hand, which renders Nazi salutes and tries to strangle him. The character is rumored to have been modeled after Edward Teller, father of the hydrogen bomb.

9. *THE ASTRO ZOMBIES*

This 1968 bad-film classic was directed by schlockmeister T. V. Mikels. John Carradine plays Dr. Marco, a mad scientist who turns people into zombies in his basement.

10. *RE-ANIMATOR*

This 1985 modern horror classic, directed by Stuart Gordon, is based on an H. P. Lovecraft story. Herbert West (played by Jeffrey Combs) is a medical student who conducts experiments to bring the dead to life. He proves too successful for his own good. A famous scene features a detached reanimated head.

Hoaxes and Deceits

A hoax perpetuated in the name of science comes in various guises. A serious hoax can destroy the credibility of an entire line of scholarship, and a medical scam can wrench not only hope but also money from the unfortunate. The lesser hoax, which is conceived in jest or mere opportunism, often upends ill-conceived theories or solipsistic attitudes that need to be revised or axed.

1. LYING STONES OF DR. BERINGER

Dr. Johann Beringer was dean of the faculty of medicine at the University of Wurtzburg in Germany. He was also an amateur naturalist. In May 1725, Dr. Beringer hired three local boys to bring him any interesting objects that they could find at nearby Mount Eivelstadt. Junior members of the faculty, annoyed by the professor's overbearing manner, hired the boys to bring fake fossils to the professor. The hoaxers tried to make Beringer realize what was happening, but he continued to write a treatise about the mysterious stones. However, when he received one bearing his own name, he realized that he had been duped and sued his fellow professors. His book, unfortunately, had already been published. *Lithographiae Wirceburgensis* was translated into English in 1963.

2. WINGED MOON CREATURES

When circulation sagged for the *New York Sun* in 1835, editorialist Richard Adams Locke concocted a series of stories about British astronomer Sir John Herschel's sightings of winged creatures flying about the moon's surface. The creatures were reportedly half man, half bat. Locke quoted Herschel describing the creatures: "The face, which was of a yellowish flesh color, was a slight improvement upon that of a large orangutan... Lieutenant Drummond said they would look as well on a parade ground as some of the old Cockney militia." Although the *Sun*'s circulation shot up, scientists were suspicious, because no telescope at the time could pick up such details. Finally, later that year, the newspaper's publisher confessed that the sighting was a hoax.

3. THE CALAVERAS SKULL

In 1866, a skull found by miners on Bald Mountain in Calaveras County, California, was presented to J. D. Whitney, state geologist and professor of geology at Harvard. When Whitney announced his discovery, he claimed that it proved the existence of Pliocene-age man in North America, the oldest known record of human existence to date. However, it was discovered that the miners had played a joke on the reserved professor by presenting him with a Native American skull dug up from a nearby burial site. Nevertheless, Whitney continued to believe that the skull was genuine.

4. PILTDOWN

Amateur geologist Charles Dawson announced in 1912 that he had found what appeared to be the partly apelike skull of an ancient human in a gravel bed on Piltdown Common near Lewes, England. However, in 1953, chemical analysis of the remains showed that the scientific establishment had been duped for four decades. The cranium belonged to a modern human and the jaw to an orangutan, skillfully stained and altered to look old. Since then, nearly everyone connected with Piltdown has been implicated in the hoax. In May 1996, paleontologist Brian Gardiner of King's College, London, and his colleague Andrew Currant finally revealed that Martin A. C. Hinton, keeper of zoology at the British Museum from 1936 to 1945, perpetrated the crime. They claim that his motivation was to embarrass Arthur Smith Woodward, keeper of paleontology at the British Museum. Apparently, Hinton had annoyed Woodward by requesting a salary and was dismissed.

5. THE TASADAY PEOPLE

In 1972, Stone Age people were discovered in the Philippines. *National Geographic* ran a cover story on the discovery, but it turned out to be a hoax. Manual Elizalde Jr., a Philippine cultural minister, announced that he had discovered in the rain forest of the Philippine island Mindanao a tribe he called the Tasadays, who had no contact with the outer world and still used Stone Age implements. However, when Swiss journalist Oswald Iten traveled there in 1986 to see the tribe, he was told that the local Tboli and Manobo peoples had been paid and provided protection to pose as primitive cave dwellers.

6. WILLIAM McBRIDE

In 1982, Australian obstetrician William McBride published a report about a morning-sickness drug called Debendox that, he claimed, caused birth defects in rabbits. Merrell Dow took the drug off the market after an avalanche of lawsuits. But there was one problem. McBride had altered data in research carried out by his assistants. The drug was proved to show no ill effects. McBride, who had previously, and correctly, warned against the dangers of thalidomine in 1961, was found guilty of scientific fraud in 1993 by a medical tribunal.

7. ROBERT SLUTSKY

Cardiac radiologist Robert Slutsky resigned in 1985 from the University of California at San Diego School of Medicine after investigators concluded that he had altered data and misrepresented methods in order to publish a new research article every 10 days. He also had persuaded prominent scientists to lend their names to his articles.

8. GALADRIEL MIRKWOOD

Polly Metzinger sought and found a deeper logic to the immune system than the self-nonself theory by realizing that a mother didn't reject a fetus and tumors didn't trigger a rejection response because there was no necrosis. However, her nontraditional thinking and behavior almost doomed her career. Refusing to write in the passive voice, and too insecure to adopt the first-person voice, Metzinger invented a coauthor—her Afghan hound named Galadriel Mirkwood—so that she could write *we*. The *Journal of Experimental Immunology* was not amused and banned her from its pages.

9. ALLAN SOKAL

Confessing that he was an "unabashed Old Leftist who never understood how deconstruction could help the working class," Allan Sokal, professor of physics at New York University, published a parody in the 1996 spring/summer issue of *Social Text,* entitled "Transgressing the Boundaries: Towards a Transformation Hermeneutics of Quantum Gravity," to see "if a leading journal of cultural studies would publish an article liberally salted with nonsense." It did.

10. MISSING LINK

In 1999, a fossil smuggled out of China that showed a dinosaur with birdlike plumage was displayed at the National Geographic Society and was documented in the society's November issue.

Some paleontologists said that the fossil was the missing link that proved birds evolved from dinosaurs. Unfortunately, it was discovered that a Chinese farmer had pieced together bits of birds' bones and a carnivore's tail.

Noted Feuds and Controversies

Scientists pride themselves on dispassionate examination of data. Yet the process of scientific discovery is fraught with emotion. When introducing a new idea, one scientist usually steps on the toes of another who doesn't give up a cherished theory easily. Some questions are resolved, or at least argued, politely, as in the case of Charles Darwin and Alfred Wallace, whereas others involve bitter recriminations, as in the battle between Newton and Leibniz. New ideas threaten established beliefs and thus evoke belligerence, blindness, and pettiness—proving that even scientists are human.

1. GALILEO VS. POPE URBAN VIII AND THE WORLD

On June 22, 1633, Galileo Galilei (1564–1642) was put on trial at Inquisition headquarters in Rome. As a cardinal, Urban had interceded on Galileo's behalf during an earlier confrontation in 1616. Although Galileo was not an atheist or religious brigand, he was forced by threat of torture and burning to "abjure, curse, and detest" his belief that the Sun, not Earth as the church taught, was the center of the universe and that the Sun moved around Earth, not vice versa. Galileo had made the mistake of writing *Dialogue on the Two Chief World Systems, Ptolemaic and Copernican* in the vernacular

Italian, instead of Latin, and presenting it as a dialogue among Sagredo (the moderator), Salviati (Galileo's voice), and Simplicio (the Ptolemaic opposition). Unfortunately for Galileo, Urban's standard argument—that God's infinite wisdom and power cannot be limited to one particular conjecture—was voiced by Simplicio, and Urban took offense to the personal insult. Galileo's life was spared, but he was put under house arrest for the rest of his life.

2. NEWTON VS. LEIBNIZ

Calculus was developed almost simultaneously by Sir Isaac Newton (1642–1727) and Gottfried Wilhelm Leibniz (1646–1716), although each used different symbols and notations. Leibniz's notations were considered superior and were preferred over Newton's, causing a bitter controversy. Their conflict quickly became a matter of national pride, with English scientists refusing to accept Liebniz's version.

The feud derived from differences in philosophical and religious ideas. To Newton, space and time were absolute entities, existing independently of the human mind. This certainty provided a foundation for classical physics. Leibniz, however, argued that if they were absolute, time and space would be independent of God, who could not control them. For him, space and time were relative concepts, not absolutes. Leibniz's theories didn't come into their own until the rise of relativity.

Who won? Newton, hands down. He was knighted and buried in Westminster Abbey. Leibniz, according to a friend, was buried like a robber rather than what he was, "the ornament of his country."

3. NEWTON VS. HOOKE

In 1672, Sir Isaac Newton was elected to the Royal Society when he reported on his experiments with light and its spectrum. While most members were impressed, physicist Robert Hooke (1635–1703), who had performed similar experiments and drawn different conclusions, dismissed Newton's findings as irrelevant. In response, Newton, who abhorred criticism, sent a scathing letter to Hooke. The feud between the two escalated, although Newton's eminence forced them to act civilly in public. Their enmity culminated in 1687, when Newton published his theory of universal gravitation. Hooke, who had outlined a similar theory seven years earlier, insisted that Newton had plagiarized his idea. However, Newton had developed his theory in the 1660s, when he was in his twenties. He published it in the 1680s only at the urging of Edmund Halley, of Halley's Comet fame.

4. LAVOISIER VS. MARAT

When the French Revolution broke out in 1789, Antoine-Laurent Lavoisier's (1743–1794) position in the government's tax collection agency made him a target of hatred. When he was charged with treason, his age-old enemy, Jean-Paul Marat (1743–1793), accused the chemist of diluting commercial tobacco and cutting off Paris's air supply by building a defensive wall around the city. On May 8, 1794, Lavoisier was guillotined.

Marat had acquired a medical education, and for some years was a successful physician in England and France. He also conducted scientific experiments in optics and electricity. Lavoisier had prevented Marat from joining the Academy of Sciences after Marat wrote a poorly documented paper concerning his observations of fire. Failure to achieve the recognition he thought he deserved left Marat feeling persecuted. He turned his energy toward fighting royal despotism and stoked the outrage felt by the downtrodden, which continued the French Revolution.

5. TESLA VS. EDISON

Thomas Alva Edison (1847–1931) was committed to the use of direct-current (DC) electricity—with no competition. However,

Nikola Tesla (1856–1943), the Serbian who perfected the efficient use of alternating-current (AC) electricity, came to the United States in 1884 and worked in Edison's research laboratory. Tesla had worked to overcome the problems of DC motors at the Continental Edison Company in Paris. When Tesla explained to Edison his plans for a motor based on AC, he created the foundation for a long feud. Furthermore, Edison asked Tesla to increase the efficiency of his dynamos for $500,000, which Edison refused to pay. Striking out on his own, Tesla established his own electric company and acquired several patents. George Westinghouse bought a patent for a polyphase motor and hired Tesla to work in his Pittsburgh plant.

Edison did everything in his power to convince the public that DC electricity was the wave of the future. He paid schoolboys a quarter for each pet that they stole for him. Using AC, Edison had the animals electrocuted and displayed to scare people about AC. One of Edison's assistants went on the road electrocuting, or as Edison put it, Westinghousing, calves and dogs.

In 1943, Tesla died a poor man, having lived his last years almost entirely on milk and Nabisco crackers, although he had been a millionaire in the 1890s. But he won in the end: AC is now the standard for electrical output.

6. **COPE VS. MARSH**

Paleontologists Edward Drinker Cope (1840–1897) and Othniel Charles Marsh (1831–1899) were men of independent means, ambitious, driven, competitive, and not overly scrupulous. They flourished during a time when antebellum America was building a transcontinental railroad and, in the process, unearthing strange bones. Darwin's *Origin of Species*, published in 1859, had fired the flames of controversy in paleontology—dinosaurs could either prove or disprove evolution. Cope, of the University of Pennsylvania, claimed that he had introduced Marsh to many dinosaur-rich excavation sites but was rewarded by having his own "areas" bought out from under him. Marsh, professor of paleontology at Yale University, funded his expeditions through government help, wealthy relatives, and personal connections, while Cope gambled on unfortunate mining ventures to compete with Marsh's funds. Both began purchasing fossils that had already been collected to keep up with each other. On January 12, 1890, the *New York Herald* announced that Cope charged Marsh with plagiarism (Cope has published voluminously) and that Marsh's assistants had smashed fossils to prevent others from getting them. Marsh countered by accusing Cope of stealing his fossils, sneaking into his workrooms, and being mentally unbalanced. Although their bitter rivalry was never settled, their efforts in collecting an astonishing amount of dinosaur fossils displayed around the country fueled interest in dinosaurs. Cope and Marsh named nearly 130 species, including Tyrannosaurus, Bracchiosaurus, and Triceratops.

7. **FREUD VS. JUNG**

Controversy still surrounds the theories of Sigmund Freud (1856–1939), but it's beyond question that the explorer of the psyche was the major architect of modern psychology. He theorized that most mental activity occurs outside of conscious awareness, that much of human suffering is the result of unconscious conflicts, and that there is a resistance to recognizing this process. Freud also showed that the cornerstone of any treatment works with that resistance and the unconscious conflict.

In 1907, Freud met Carl Gustav Jung (1875–1961) in Vienna. The enthusiastic Jung had read Freud's *Interpretation of Dreams,* and Freud saw Jung as his spiritual son who could ensure the survival of psychoanalysis. Jung was not only brilliant, but he was also not Jewish, as were Freud's other followers, and thus had a better chance of gaining scientific credibility in the anti-Semitic climate of Vienna.

Sigmund Freud, considered by many to be the founder of modern psychology, broke with his adherent Carl Jung over the place of religion in psychoanalysis. Eventually Freud came to believe that Jung had a death wish for him.

Although Jung became the first president of the International Association of Psychoanalysis in 1910, he resigned in 1914 to found his own movement that embraced spirituality, which was most unwelcome to Freud, an aggressive atheist.

Perhaps Freud perceived the possibility of this break earlier, when he and Jung had lunch together before boarding a ship to travel to the United States for a lecture series in 1907. Jung couldn't stop talking about an archaeological dig. Freud became alarmed because he considered this proof of Jung's death wish against him, and Freud fainted. In 1912, Jung detailed his "heretical" views in a lecture series in the United States, and Freud fainted again in front of Jung while discussing how Jung had published his views without citing Freud's work. Jung then sent angry letters to a bewildered Freud. After Jung announced his departure from Freud's circle, Freud again fainted. After the break, Jung suffered a period of intense anxiety during which he could not work effectively.

8. **OPPENHEIMER VS. TELLER**

Even though he brought forth a weapon of mass destruction, theoretical physicist J. Robert Oppenheimer (1904–1967) possessed charisma that drew top scientists into the Manhattan Project to build an atomic bomb. Edward Teller (1908–) was involved in the project and had lobbied President Franklin D. Roosevelt to build atomic bombs. One of Teller's tasks was to calculate the possibility of developing a hydrogen bomb. Once he understood that it was possible, he devoted much energy to his campaign for building a hydrogen bomb, even after World War II.

When the Cold War began, and the Russians exploded their own atomic bomb in 1949, the development of the H-bomb became top priority. Teller lobbied and received the go-ahead to establish a second nuclear weapons laboratory at the Lawrence Livermore Laboratory. Oppenheimer opposed the development of more destructive nuclear weapons. This caused such resentment among government circles that the Atomic Energy Commission filed charges against Oppenheimer based on his association with communists (including his wife). Teller testified against him, saying that Oppenheimer's actions "appeared confused and complicated" and that he would prefer to see vital security interests entrusted to other hands. Although the committee did not find Oppenheimer disloyal, it did revoke his security clearance.

Afterward, Teller was shunned by the scientific community, but he continued to exert power even in the Reagan administration. The character of Dr. Strangelove of Stanley Kubrick's brilliant film of the same name was based on Teller's views and handicap (his foot was severed in a streetcar accident, and he walks on a prosthetic foot but doesn't use a wheelchair). Oppenheimer received the prestigious Enrico Fermi Award in 1963, but he had been deeply wounded by the whole affair.

9. FREEMAN VS. MEAD

American anthropologist Margaret Mead (1901–1978) redefined and expanded anthropology. She popularized the idea that human differences arose from cultural influences at least as much as from biological determinants. Her first publication (she eventually published two dozen books), *Coming of Age in Samoa,* was published in 1928, and the material in the book was collected when she was in her early twenties.

Mead studied Samoan adolescents to see if they experienced the same turmoil as their American counterparts. She concluded that Western civilization's family organization crippled emotional life and warped individual growth. She said that Samoans condoned adolescent free love, which helped smooth the transition from adolescence to adulthood.

By the time of her death in 1978, Mead's reputation seemed secure. However, in 1983, the *New York Times* featured a review of Derek Freeman's book, *Margaret Mead and Samoa: The Making of an*

Anthropological Myth. Freeman, professor emeritus at Austrian National University, had studied Western Samoan culture for years. He claimed that Mead had deceived herself by her belief in the "nurture" theory, and that her assumptions were fundamentally wrong and unsubstantiated. In 1989, he also claimed that Fa'Apua'a, one of the young Samoans Mead had interviewed, had lied about free sex on the island. The controversy has sullied Mead's reputation, yet she remains one of the most remarkable anthropologists of the twentieth century.

10. JOHANSEN VS. LEAKEY

Louis (1903–1972) and Mary (1913–1996) Leakey zeroed in on the Olduvai Gorge as their site to prove that Darwin was right—humans originated in Africa. In the summer of 1959, after 30 years of searching, Mary unearthed a skull with huge jaws and a gorilla-like crest, yet with certain human characteristics. Louis labeled it *Zinjanthropus boisie,* the earliest human ancestor and, in fact, the missing link (it was later labeled in the Australopithecus group). The find put the Leakeys on the map.

At the same time, their son Richard (b. 1944) was growing up, learning the craft in the field. He focused on a different region, Koobi Fora, Kenya, north of Olduvai. In 1972, Richard presented his father with a skull with a large braincase and without the prominent brow. He believed that this was a true human ancestor who had lived two or more million years ago. In 1973, when Richard published his find, he named it *Homo habilis,* the toolmaker. By 1979, at age 35, Richard had collected as many important finds as his father had in the previous 30 years.

Meanwhile, Donald Johanson (b. 1943) was working on his doctoral dissertation in 1970. A geologist wishing to study the plate tectonics of the Afar triangle, at the northern end of the Great Rift Valley, wanted to take a paleontologist on a field trip there. Richard recommended that he take Johanson, which he did. Johanson became obsessed with finding hominids. In the fall of 1974, he found a skeleton dated more than three million years old and nicknamed it "Lucy." The find rocketed him to the same eminence as Richard. A colleague, Tim White, convinced Johanson that Lucy represented a new species, which they named *Australopithecus afarensis,* and that she was bipedal. Accounts of the discovery claim that Johanson called out, "Hey, Richard, look at this one. I've got you now, Richard."

Richard was not pleased, especially with Johanson's claim that his finding had pushed back the origins of humans by 1.5 million years. At the official announcement at a Nobel symposium in 1978, Johanson announced that he was including a number of Mary's fossil finds in his classification. The feud was later exacerbated by Johanson when he published his findings as the true missing link.

The feud culminated on television in 1981, when Walter Cronkite invited Johanson and Richard Leakey to discuss the situation. Richard felt that he had been trapped by a better-prepared Johanson and drew a big *X* across the diagram Johanson drew of the human tree.

The two haven't spoken since. Richard withdrew from the field. In 1993, he lost both legs when his plane crashed.

People Who Gave Their Lives and Limbs to Science

Experimentation is at the root of scientific discoveries, but it is also the cause of many injuries and, in some cases, death.

1. GIORDANO BRUNO

After being imprisoned for seven years, Italian monk Giordano Bruno (1548–1600) was burned at the stake in 1600 by the Inquisition in part for his heretical teaching throughout Europe that Earth revolved around the Sun and that there may be an infinite number of Earthlike worlds and suns.

2. ANTOINE-LAURENT LAVOISIER

The father of modern chemistry, Antoine-Laurent Lavoisier (1743–1794) was the first scientist to explain how things burn and developed the first rational system for naming chemical compounds. Although he was fairly wealthy, the French chemist earned little money and was forced to invest his funds in a private tax-collection service for the government. With the outbreak of the French Revolution in 1789, his position as an administrator in a tax-collection agency made him a target of hatred. He was barred from his laboratory and arrested. He protested that he was a scientist, but he was

Pioneer chemist Antoine-Laurent Lavoisier, shown here with his wife, was the first to explain how things burned and developed the first rational system for naming chemical compounds. However, his position in a tax-collection agency made him a target of hatred during the French Revolution, and he was guillotined on May 8, 1794.

told, "The Republic has no need of scientists." An old enemy, Jean-Paul Marat (1743–1793), accused Lavoisier of diluting commercial tobacco and cutting off Paris's air supply by building a defensive wall around the city. On May 8, 1794, Lavoisier was guillotined.

3. KARL WILHELM SCHEELE

A Swedish chemist, Karl Wilhelm Scheele (1742–1786) discovered many chemical elements, often through sniffing or tasting his discoveries, even hydrogen cyanide. He died from symptoms resembling mercury poisoning.

4. SIR HUMPHRY DAVY

The inventor of the electrolytic technique of chemical analysis, English chemist Sir Humphry Davy (1778–1829) was a victim of his habit of inhaling gases. Even though he discovered the anesthetic properties of nitrous oxide, chemical poisoning left him an invalid the last 20 years of his life. He also damaged his eyes in a nitrogen chloride explosion in 1812.

5. MICHAEL FARADAY

English chemist and physicist Michael Faraday (1791–1867) benefited from Sir Humphry Davy's eye injury in that he became his secretary and protégé. He went on to make seminal discoveries in electromagnetics, but he was plagued by an eye injury sustained in another nitrogen chloride explosion and later suffered from chronic chemical poisoning.

6. MARIE CURIE

Polish-born French chemist Marie Curie (1867–1934) devoted her life to radiation research after the death of her husband, Pierre, in 1906. She contracted leukemia from overexposure to radioactivity and suffered a painful death in 1934.

7. GERTY RADNITZ CORI

Just after Gerty (1896–1957) and Carl Cori (1896–1984) won the Nobel Prize in 1947 for discovering the enzymes that convert glycogen into sugar and back to glycogen, they learned that Gerty was suffering from agnogenic myeloid dysplasia, a fatal type of anemia. It is likely that her illness was triggered by the X rays that she conducted while studying the skin and metabolism of body organs.

8. EMIL FISCHER

Organic chemist Emil Hermann Fischer (1852–1919) won the 1902 Nobel Prize in chemistry, but he wasn't aware of the toxicity of his discovery, phenylhydrazine. He contracted chronic poisoning from its continual use. When a physician told him that he was suffering from serious intestinal disorders, Fischer thought he had cancer and ended his life with cyanide.

9. EUGENE SHOEMAKER

On July 18, 1997, geologist Eugene Shoemaker was killed in a head-on automobile collision while on an annual trip to the Australian outback to study sites where he believed asteroids may have struck Earth. He had first studied the Barringer Crater in Arizona, where he found and identified the mineral coesite, which is created by a powerful explosion. After showing that impacts from space can cause craters on Earth, he turned his sights toward the Moon. His bid to become an astronaut was foiled by a medical condition, but he instructed the *Apollo* astronauts in geology and studied the rocks that they brought back from the Moon. After he died, his ashes were sent to the Moon on the lunar *Prospector* in 1998.

The comet Shoemaker Levi 9 impinging on Jupiter in 1997 was named after Shoemaker and his wife, who systematically scanned sky photos for years and discovered many new stars and comets.

The comet gave astronomers more information about Jupiter's atmosphere and state than they had before.

10. **DR. HEINZ PAGELS**

Each summer, scientists arrive at the Aspen Center for Physics—three low-slung buildings on the outskirts of the ski resort in Colorado—to discuss and write about physics while hiking and cycling up mountain passes. In 1988, Dr. Heinz Pagels, a Rockefeller University physicist, fell to his death there—a death that he had dreamed about and described in his 1982 book, *Cosmic Code.*

Noted Women Scientists before Women's Equality

Preconceived notions about the respective natures of men and women dominated perceptions of what women could and could not do in science. Women didn't participate actively in science until the nineteenth century, when they worked as data gatherers and crystallographers. Political and educational reforms of the late nineteenth century provided increased opportunities for women's activity in science.

1. CAROLINE LUCRETIA HERSCHEL

English astronomer Caroline Lucretia Herschel (1750–1848) helped her brother William (1738–1822) catapult to fame in 1781 through his discovery of a new "comet," later recognized as Uranus. At first, Caroline helped her brother grind and polish mirrors and copied catalogs, but she proceeded to discover eight comets (five of which are credited to her) over the period 1786 to 1797. In 1787, she was granted a salary by the king as William's assistant. She also received a gold medal from the Royal Astronomical Society in 1828 for her catalog of nebulae, which proved indispensable to William's son John.

2. MARY ANNING

British paleontologist Mary Anning (1799–1847) inherited her father's love of fossil collecting. Although she lacked formal education, she made several important finds, including the first complete icthyosaur skeleton in 1818.

3. MARIE SKLODOWSKA CURIE

Polish physicist and radiation chemist Marie Curie (1867–1934) was the first woman to win a Nobel Prize and one of few scientists ever to win that award twice. In collaboration with her husband, Pierre, Marie developed and introduced the concept of radioactivity to the world. In Warsaw, she entered the "Floating University," an underground, revolutionary Polish school that prepared young Polish students to become teachers. She and her sister Bronya sought education abroad, but the family couldn't afford to send either of them, so Marie took a job as a governess to fund her sister's medical education in Paris. In 1891, Curie enrolled at the Sorbonne and became one of the few women in attendance at the university. In 1893, she received a degree in physics, finishing first in her class. The following year, she received her master's.

After her marriage to Pierre and the birth of their daughter Irene, Marie began to work on her doctorate. Fascinated by the work of French physicist Antoine-Henri Becquerel (1852–1908) concerning rays emitted from uranium, she tested all the known elements to see if any, like uranium, caused the nearby air to conduct electricity. In her studies, she found that additional radioactive elements must cause the minerals pitchblende and chalcolite to emit more rays than could be accounted for normally. In 1898, she and Pierre extracted an element—polonium—from the ore that was even more radioactive than uranium. Six months later, they discovered radium embedded in pitchblende. In 1903, Marie became the first woman to complete her doctorate in France, summa cum laude.

After Pierre was killed while walking in the congested streets of Paris, Marie assumed his teaching position at the University of Paris, thereby becoming the first woman to receive a post in higher education in France, although she wasn't named to a full professorship for two more years. During World War I, she instructed army medical personnel in the practical applications of radiology. She saw her daughter Irene receive a doctorate, but, as a result of prolonged exposure to radium, Marie contracted leukemia and died in a nursing home in the French Alps. Marie was buried next to her beloved Pierre in Sceaux, France.

4. ANNIE JUMP CANNON

American astronomer Annie Jump Cannon (1863–1941) graduated from Wellesley College in 1884 and afterward went deaf as a result of contracting scarlet fever. She went on to study astronomy at Radcliffe College and became a protégée of Edward Pickering at Harvard College Observatory, where she became a calculator and curator of the astronomical photo library. Cannon's lasting legacy is the classification system for stars that is still used today. She revised the system Pickering established, basing hers on surface temperature, and therefore color. In order of decreasing surface temperature, this produced the system O, B, A, F, G, K, M, R, N, and S.

5. HELEN DEAN KING

U.S. biologist Helen Dean King (1869–1955) stirred a heated public debate when she concluded from her studies of inbred rats that the animals compared favorably with stock albinos. Reporters implied that she considered incest taboos unnecessary, whereas she asserted that certain benefits result from both inbreeding and outbreeding. This work, plus her domestication of the Norway rat, helped make possible the maintenance of pure strains of laboratory animals.

6. MARIA MITCHELL

U.S. astronomer Maria Mitchell (1818–1889) absorbed a love of astronomy from her Quaker father, William Mitchell, who spent evenings observing the heavens. Maria learned to operate a sextant and a clumsy reflecting telescope. On October 1847, she observed a new comet but didn't receive credit for it until a year later because her letter of discovery wasn't posted quickly enough. She was elected the first woman member of the American Academy of Arts and Sciences in 1848 and the American Association of the Advancement of Science in 1850. In 1865, Mitchell became professor of astronomy and director of the observatory at the newly founded Vassar College.

7. NETTIE MARIA STEVENS

U.S. cytogeneticist Nettie Maria Stevens (1861–1912) was one of the first American women to be recognized for contributions to scientific research. She discovered the chromosomal determinants of sex. Although the behavior of chromosomes had been explained, no link had been made for a chromosomal basis for heredity. The breakthrough resulted from her study of the common meal worm.

8. HERTHA MARKS AYRTON

Hertha Marks Ayrton (1854–1923), born Phoebe Sarah Marks, invented and patented an instrument for dividing a line into equal parts in 1884. She attended Finsburg Technical College, where she met W. E. Ayrton, professor of physics, whom she married in 1885. In the 1890s, she experimented with electricity and published a book, *The Electric Arc,* in 1902, which became the accepted textbook on the subject. In 1904, when she read a paper before the Royal Society, on the causes of ripple marks in sand, she was the first woman to do so. Although Ayrton couldn't become a fellow of the society, she received the Hughes Medal for original research in 1904.

9. SOPHIA JEX-BLAKE

The courage and tenacity of Sophia Jex-Blake (1840–1912) opened medical training in British schools to women. Her book, *Medical Women* (1886), chronicled her fight with the University of Edinburgh to allow women to attend regular classes and to be admitted to Surgeon's Hall so that they could observe surgeries and testing. In 1877, she was granted the right to practice medicine in Great Britain by the Irish College of Physicians. In 1878, she began practicing in Edinburgh, where she organized a school of medicine for women.

10. ELIZABETH GARRETT ANDERSON

Physician-surgeon Elizabeth Garrett Anderson (1836–1917) opened the medical profession to women in England. As a surgical nurse, she followed doctors around the wards to gain her medical education. The universities of London, Edinburgh, and St. Andrews refused to allow Anderson to earn an M.D. degree, but she gained licensing through the Apothecaries Hall in 1865. She became a physician for women and children in London. In 1868, it became possible for women to obtain M.D. degrees in France, and Anderson received a degree in 1870. She then was accepted as the visiting medical officer at Middlesex Hospital.

Eminent Scientists Who Didn't Win a Nobel Prize

From the Nobel Prize's inception in 1901, any scientist was eligible for the award in physics, physiology, or medicine. Almost all of the following were nominated but failed to win.

1. DMITRI MENDELEEV

Russian chemist Dmitri Mendeleev (1834–1907) formulated the Periodic Law of the elements, which proved that related elements appear at regular intervals, thus predicting unknown elements.

2. JOSEPH LISTER

Inspired by Louis Pasteur's research concerning the transmission of microorganisms by air, Joseph Lister (1827–1912) introduced antiseptics, making possible later advances in surgery.

3. PIERRE-EUGÈNE-MARCELIN BERTHELOT

Repeatedly nominated for the Nobel Prize, French chemist Pierre-Eugène-Marcelin Berthelot (1827–1907) made important contributions to organic and physical chemistry. He also published alchemical works and laid the foundation for chemical archaeology.

4. FRITZ RICHARD SCHAUDINN

German zoologist Fritz Richard Schaudinn (1871–1906) discovered the causal organism of syphilis with Erich Hoffman in 1905. Schaudinn was nominated for the Nobel Prize, while Hoffman was not.

5. JOSIAH WILLARD GIBBS

American chemist Josiah Willard Gibbs (1839–1903) was a pioneer in the field of chemical thermodynamics (the application of mathematics to chemical subjects), laying the groundwork for modern physical chemistry.

6. WILLIAM WALLACE CAMPBELL

One of the most influential astronomers of the twentieth century, the American William Wallace Campbell (1862–1938) determined stars' velocities in relation to Earth's and the Sun's motions, laying the foundation for astrophysics.

7. THOMAS ALVA EDISON

Thomas Alva Edison (1847–1931) obtained more than 1,000 patents for inventions, including the electric light, phonograph, telegraph, and talking motion pictures. He was nominated for a physics prize for his discovery of thermionic emission, or the "Edison effect," which established that a vacuum lamp permits electric currents to pass into a vacuum.

8. ERNEST HENRY STARLING

British physiologist Ernest Henry Starling (1866–1927) was a pioneer in the study of the heart's pumping action. With Sir William Bayliss, he coined the term *hormone,* as well as identifying secretin, which is released by the pancreas. Among his other accomplishments were studies of the circulatory system and the secretion of lymph and other body fluids.

9. ALBERT SABIN

American virologist Albert Sabin (1906–1993) developed an oral vaccine that immunized against the infection and paralysis of polio.

10. JONAS SALK

American microbiologist Jonas Salk (1914–1995) developed and tested publicly in 1954 the first vaccine to immunize against the paralysis of polio. Sabin improved on his work five years later.

Names That Became Scientific Terms

The following scientists, whose names are used as scientific measurements, are known as giants in their own fields.

1. NEWTON

Named after Sir Isaac Newton (1642–1727), a Newton equals the force with which a mass of 101.97 grams presses downward under standard conditions. It is defined as the inertia of mass or the force needed to accelerate a given mass ($F = m \times a$), the second of Newton's axioms. Because Earth is not exactly spherical, gravity is different in different places on Earth.

In the *Principia Mathematica* (1687), Newton arranged Galileo's findings into three basic laws of motion: (1) a body at rest remains at rest, and a body with a uniform velocity remains in motion, unless acted on by an external force; (2) a force is equal to mass times acceleration; and (3) for every action, there is an equal and opposite reaction. Newton's inverse-square law of gravitation allowed him to quantify gravitational attraction between Earth and the Moon. He also estimated rather accurately the masses of Jupiter and Saturn.

2. WATT

The watt, a measurement of the rate of energy conversion, is named after James Watt (1736–1819), the Scottish engineer who made possible the Industrial Revolution with his work on the steam engine. He also applied a centrifugal governor to the steam engine, not to control steam output but to keep engine speed constant under different loads.

3. VOLT

The volt, a unit of electrical potential difference, is named after Allesandro Giuseppe Antonio Anastasio Volta (1745–1827), an Italian physicist who devised the world's first electric battery and the voltaic pile, a stack of round copper and zinc disks and cotton cloth dipped in diluted sulfuric acid. The volta column had a short lifetime because it was a "wet" column, but it could produce a powerful electrical current.

4. FARAD

The farad, a unit of electrical capacitance, is named after English physicist Michael Faraday (1791–1867), who discovered benzene and the principle of current induction. One of his first triumphs was to produce continuous rotation by the use of electric current, thus making possible the electric motor. In 1831, he discovered electromagnetic induction and invented the voltameter, which measures electric charge.

5. KELVIN

The Kelvin, a unit of absolute temperature, is named after the father of thermodynamics, William Thomson (Lord Kelvin, 1824–1907). Almost every honor that can be bestowed upon a scientist was awarded to Lord Kelvin, including burial at Westminster Abbey. In 1848, he originated absolute temperature scales; in 1856, proposed

Names That Became Scientific Terms

The kelvin, a unit of absolute temperature, is named after Lord Kelvin (William Thomson, 1824–1907), the father of thermodynamics.

laws of thermodynamics; and, in 1866, engineered the first transatlantic telegraph cable.

6. MACH

Austrian physicist Ernst Mach (1838–1916) rejected Newton's concept of absolute time. He was the first to realize that matter traveling through the air moving faster than the speed of sound drastically altered the quality of the space in which it moved. Mach's name expresses the speed of matter relative to the speed of sound at a certain temperature.

7. TESLA

The tesla, a unit of magnetic flux density, is named after the Serbian Nikola Tesla (1856–1943), the eccentric inventor who proved and perfected the efficient use of the alternating current.

8. HERTZ

This unit of frequency is named after German physicist Heinrich Rudolf Hertz (1857–1894). Hertz is best known for his work with electromagnetism and electromagnetic radiation, or the radio wave. A range of frequencies in which Hertz experimented (2 inches to 2 feet) is called "Hertz' waves" even today. Hertz, together with Guglielmo Marconi (1874–1937), used the distant effect of electromagntic waves for information transmission, and then Marconi built the first wireless telegraphs.

9. GUTENBURG DISCONTINUITY

The refraction of seismic waves demonstrates that Earth has a core. Seismic waves reaching a depth of 2,900 kilometers suddenly encounter a new substance that is much denser than the overlying rock. This boundary, discovered by German-born U.S. geophysicist Beno Gutenberg (1889–1960) in 1914, is called the Gutenberg Discontinuity. The boundary causes some energy to be reflected back to the surface.

10. FULLERENE

American inventor and engineer Buckminster Fuller (1895–1983) popularized the geodesic dome. The fullerene is a large molecule containing 60 carbon atoms arranged in a soccer ball–shaped cage. The molecules are the rarest form of naturally occurring carbon on Earth.

Creative Couples

Scientific breakthroughs often occur through the collaborative efforts of partners. Here we examine the efforts of men and women who contributed to the advancement of science together.

1. THE HUGGINSES

William Huggins (1824–1910) was one of the founding fathers of stellar spectroscopy. Margaret Huggins (1854–1915) was more than her husband's able assistant, secretary, and illustrator. She was a strong impetus behind his photographic research, a painstaking activity at the time.

2. THE JACOBIS

Mary Putnam Jacobi (1842–1906) and Abraham Jacobi (1830–1919) proved that physicians could not only marry but unite their strengths to provide more service to more people. Called the future of America's pediatrics, Abraham was a major figure behind the development of Mount Sinai Hospital and the New York Academy of Medicine. Mary worked when women physicians were barred from inpatient care, through the death of her firstborn from diphtheria (the disease for which Abraham was known as an authority), and a brain tumor, which resulted in her death.

3. THE COMSTOCKS

John Henry Comstock (1849–1931) and Anna Botsford Comstock (1854–1930) forged new paths in nineteenth-century natural history. John Henry, chair of the Department of Entomology at Cornell University, was noted for his research on the evolution of insects and his innovative teaching methods. Anna dropped out of college before she married but finished her degree in science and received training as a scientific illustrator. Using these skills and her love of nature, Anna became a leading figure in the nature-study movement of the 1890s.

4. MARIE AND PIERRE CURIE

Chemist Marie Sklodowska Curie (1867–1934) was consistently recognized as her physicist husband Pierre's (1859–1906) collaborator. When they received the Nobel Prize in physics in 1903, they were presented as the "best illustration of the old proverb 'union is strength.'"

5. CARL AND GERTY CORI

Carl Cori (1896–1984) and Gerty Radnitz (1896–1957) met in their first year of medical school at the Carl Ferdinand University in Prague. They published their first joint research project in 1920 on immune bodies in blood in various diseases. Upon receiving the Nobel Prize in physiology and medicine in 1949, Carl said, "Our efforts have been largely complementary, and one without the other would not have gone as far as the combination."

6. IRÈNE AND FRÉDÉRIC JOLIOT-CURIE

Irène Curie (1897–1956) was her mother's research associate at the Radium Institute in Paris. There she met her husband, Frédéric Joliot (1900–1958). They began their collaboration in 1930, which culminated in the joint discovery of artificial radioactivity in 1934, for which they were awarded the Nobel Prize for chemistry in 1935.

7. **THE GAPOSCHKINS**

Astronomers Cecilia Payne-Gaposchkin (1900–1979) and Sergei Gaposchkin (1898–1984) accurately related spectral classes of stars to the actual temperature by accounting for the amounts of ionized states of an element as shown by their relative line strengths. Cecilia's major research was on the spectra of large, luminous stars and variable stars. She married Sergei in 1934, and they collaborated on studies of variable stars. Although she spent her entire career at Harvard, she was given the title "astronomer" only in 1938 and her professorship in 1956.

8. **THE LONSDALES**

Kathleen Yardley Lonsdale (1903–1971) and Thomas Lonsdale (1905–1974) combined research and practical science. During their marriage, Thomas earned the major income for the couple as a scientist in the silk industry and later as an engineer. As Kathleen's reputation as a crystallographer rose, Thomas retired early to assist her in promoting peace and prison reform. Among Kathleen's accomplishments were the demonstrations of the planarity of the benzene ring and the reality of sigma and pi electrons, as well as her groundbreaking work on the magnetic anisotropy of crystals. She was the second woman to become a fellow of the Royal Society and the first woman president of the British Association for the Advancement of Science.

9. **THE LEAKEYS**

Louis (1903–1972) and Mary (1913–1996) Leakey made discoveries that revolutionized thought on the origins of man. Their names were not always linked. Despite his teachers' insistence that life originated in Asia, Louis concentrated on east Africa, particularly Olduvai Gorge, a 25-mile-long, dry canyon that slices through what was formerly a lake. During one of his stays in England in 1933, Louis met Mary Douglas Nicol, who served as an assistant to an archaeologist in southern Britain. She illustrated stone tools for publications, and Louis asked her to contribute illustrations for his book *Adam's Ancestors*. Despite a 10-year age gap and the fact that Louis was married, they became constant companions. In 1936, they married. Mary discovered the Proconsul and Zinjanthropus (later named Australopithicus) fossils, as well as footprints in Tanzania.

10. **GEORGE PALADE AND MARILYN FARQUHAR**

Cell biologists George Palade (b. 1912) and Marilyn Farquhar (b. 1928) demonstrated by electron micrography that molecules and ions are engorged by sacs or vesicles that move to the surface from within a cell. The vesicles, named lysosomes, merge with the outer membrane, then swallow up and bring the substances inside the cell. Before their discovery, it was thought that molecules simply entered the cell through pores in the membrane. Following the death of his wife in 1969, Palade married Farquhar in 1970. In 1974, he shared the Nobel Prize in physiology or medicine for describing the structural function of the cell.

Codiscoveries

Many discoveries are so critical and challenging that they require more than one person to discover them. Sometimes scientists reach solutions almost simultaneously and independently. Whether these codiscoverers get along is another matter.

1. RUSH TO THE PUBLISHER

Although calculus was approached gradually by many people, Englishman Sir Isaac Newton (1642–1727) and German philosopher, mathematician, and logician Gottfried Wilhelm Leibniz (1646–1716) are recognized as the inventors.

Modern scholars agree that the Englishman developed calculus in 1665–1666, before Leibniz, whose work was done independently in 1673–1676. Leibniz worked independently of Newton, and he published his work before Newton. The controversy over the priority of discovery carried on by their supporters was blatantly nationalistic, and one of the longest and most bitter in the history of science.

2. OXYGEN-RICH ATMOSPHERE

Joseph Priestley (1733–1804) invented soda water, started the study of photosynthesis, and discovered oxygen. But Carl Wilhelm

Scheele (1742–1786) gets equal credit for independently discovering oxygen in Sweden just before Priestley. However, Priestley published first in a race to the finish, a trend science sees often.

3. **OPPOSITES REPEL**

Antoine-Laurent Lavoisier (1743–1794) was the opposite of Joseph Priestley in every respect. Lavoisier received a solid scientific training at the Royal Academy of France, had inherited his money, and was a part of the government establishment. In chemistry, he was a radical. Priestley ignored the Church of England; had to sell his books for money and wrote them with a commercial market in mind; and was antiestablishment. But in chemistry, he was a conservative.

Priestley told Lavoisier of his discovery of oxygen combustion. Lavoisier repeated Priestley's experiments and named the gas "oxygen." From these experiments, Lavoisier summarized his findings into the famous Law of Conservation (of matter).

Lavoisier claimed that the discovery was his, and Priestley resented him for it. While Lavoisier was guillotined during the 1794 Reign of Terror in France, Priestley sympathized with French revolutionists and was driven out of England by a mob who burned his home. Priestley was exiled to the United States and Benjamin Franklin helped him resettle, but he remained homesick until he died.

4. **ON THE ORIGIN OF ORIGINS**

Alfred Russel Wallace (1823–1913) was born in Wales, the son of middle-class English parents. Around 1835, his family fell on hard times, and Wallace was forced to move to London to work, where he decided to make his living as a natural history collector and surveyor. Eventually he did change careers to a natural history collector and traveled to the Amazon in 1848.

On his return from the Amazon, Wallace published three books and, in 1858, an essay entitled *On the Law Which Has Regulated the Introduction of New Species.* This essay established him as the codiscover (with Charles Darwin) of the theory of evolution by means of natural selection.

Weak with malaria, Wallace one day had a flash of insight on how species change. The result was his scientific paper *On the Tendency of Varieties to Depart Indefinitely from the Original Type.* Rather than send the manuscript directly to a publisher, Wallace sent it to his correspondent Darwin. Upon seeing Wallace's work, Darwin ended a 20-year delay in publishing his own theory and the next year published *On the Origin of Species* (1859).

Although Wallace independently reached similar conclusions, Darwin's name alone has usually been associated with the theory. With no resentment, Wallace commented late in life that his greatest achievement had been to prompt Darwin to publish.

5. **TIMING IS EVERYTHING**

Archibald Scott Couper (1831–1892) began work in Charles-Adolphe Wurtz's (1817–1884) laboratory, where he immediately synthesized two new carbon compounds. In 1858, the Scotsman wrote "A New Chemical Theory," on how carbon combines, and asked Wurtz to present it to the French Academy. Wurtz was not a member and delayed the release. In the interim, the German Friedrich August Kekule von Stadonitz published similar information and received credit. Couper had a nervous breakdown and spent the rest of his life as a semi-invalid.

6. **PRESENTATION IS KEY**

Jacobus van't Hoff (1852–1911), 22 years old, had a better written paper detailing the carbon atom than Joseph-Achille Le Bel

(1847–1930), 27 years old. At one time, they worked in the same laboratory but kept their ideas to themselves. Le Bel published first, but the scientific world paid no attention. Then van't Hoff expanded the idea into a book, which he sent out with cardboard models. Some liked the book and some didn't, including Herman Kolbe, who wrote such a shockingly negative review that the book attracted even more attention.

7. **FORGET ABOUT IT**

In search of a research and clinical practice post, Dr. Alois Alzheimer (1864–1915) became a research assistant to Emil Kraepelin (1856–1926) at the Munich University clinic, who was creating a new laboratory for brain research. Having published many papers on conditions and diseases of the brain, in 1906 Alzheimer gave a lecture in which he identified an "unusual disease of the cerebral cortex" that affected a female patient, Augusts D, causing memory loss, disorientation, hallucinations, and ultimately her death at age 55. After her death, Alzheimer examined her brain and found innumerable roundish molecules, which he called plaques, that were only recorded in the elderly.

Kraepelin had no choice but to name the disease after Alzheimer. Kraepelin was already known as the father of modern psychiatry and had discovered schizophrenia and manic depression.

8. **MAY THE BEST MAN WIN**

In 1922, physician and physiologist Sir Frederick Grant Banting (1891–1941) and physiologist Charles Herbert Best (1899–1978), both Canadians, codiscovered the pancreatic hormone insulin, which is used in treating diabetes. The discovery took place while they worked at the University of Toronto in the laboratory of Scottish physiologist John James Rickard Macleod (1876–1935).

However, in 1923, the Nobel Prize in medicine was awarded to Banting and Macleod. Objecting to the credit given to Macleod, who had only provided the laboratory and not participated in the discovery, Banting shared his half with Best, while Macleod divided his share with the Canadian chemist James Bertram Collip, who had helped Macleod purify insulin subsequent to its isolation.

That same year, the university established the Banting-Best Department of Medical Research with Banting as its director. In 1934, he was made Knight of the British Empire. At the height of his career, Banting died in a plane crash en route to England to take up a wartime post.

Best became a research associate of the department and director after Banting's death. During World War II, Best was influential in starting a Canadian program for the procurement and use of dried human blood serum. In 1963, Best was named adviser to the medical research committee of the United Nations World Health Organization.

9. **A BETTER MOUSETRAP**

German-American physicist Hans Dehmelt (1922–) devised the trap that isolated a single electron. As a result of his discovery, physicists revised their estimates of the electron's size by a factor of 10,000. He and his team also trapped a positron (antielectron) for three months, which they named Priscilla.

Dehmelt shared the 1989 Nobel Prize in physics with Wolfgang Pauli, who had also devised an ion trap.

10. **WATSON, CRICK, AND FRANKLIN**

Rosalind Franklin (1920–1958), while creating the world's best X-ray diffraction pictures of DNA, discovered that the sugar-phosphate backbone of DNA lies on the outside of the molecule. Maurice

Wilkins, a colleague who was also working on DNA, disliked her and showed her work to James Watson, a competitor.

Franklin died of cancer in 1958, at age 37. In 1962, the Nobel Prize went to Watson, Crick, and Wilkins. In Crick's view, if Franklin had lived, "It would have been impossible to give the prize to Maurice and not to her" because "she did the key experimental work." Franklin's notebooks show her working toward the solution until they found it; she had narrowed the structure down to some sort of double helix.

Scientists with Disabilities

Many scientists have surmounted physical problems to enrich our base of knowledge.

1. JOHANNES KEPLER

German astronomer and mathematician Johannes Kepler (1571–1630) suffered an attack of smallpox as a child. The illness left him with crippled hands and damaged eyesight. However, he went on to propose that planets orbited the Sun in elliptical paths and paved the way for Newton's fluxions and Leibniz's differential calculus.

2. JOHN GOODRICKE

A young astronomer, John Goodricke (1764–1786) discovered the variability of the star Algol. He was born a deaf-mute in Holland to an English family and was educated in England, where his best friend's father inspired him to study the heavens. At age 19, he observed that Algol, normally a second-magnitude star, was so faint that it looked like a fourth-magnitude star. He reasoned that Algol had an invisible companion that blocked the star's light when it orbited in front of it.

3. ROBERT WILHELM BUNSEN

German inorganic and physical chemist Robert Wilhelm Bunsen (1811–1899) explored the organic field with disastrous results. During a study of arsenic derivatives, he lost the sight in his right eye in a cacodylic cyanide explosion. In 1855, he went on to improve the gas burner that bears his name. Bunsen refused to take out a patent on his invention; he believed that scientists should not become wealthy as a result of discoveries.

4. ANNIE JUMP CANNON

The daughter of a prosperous shipbuilder and state senator, Annie Jump Cannon (1863–1941) was partially deaf from an early age. She rearranged the classes of stars into the familiar O, B, A, F, G, K, M sequence and made many noteworthy stellar discoveries. In 1925, Cannon became the first woman to receive an honorary degree of Doctor of Science from Oxford University.

5. CHARLES STEINMETZ

A hunchback from birth, electrical engineer and mathematician Charles Steinmetz (1865–1923) held more than 200 patents for General Electric and was a professor at Union College in Schenectady, New York. His discovery of magnetic hysteresis led to the development of energy-efficient motors. While studying at the University of Breslau, he became a socialist and ran the student socialist newspaper. With the secret police on his tail, he fled Germany for Switzerland and emigrated in 1889 to the United States, where he worked on alternating-current motors.

Because of the inherited condition that caused his deformity, Steinmetz never married or had children. He adopted an associate at General Electric and legally became a grandparent to the associate's children. He also kept a zoo of animals in his backyard and raised orchids.

6. CHARLES J. H. NICOLLE

The recipient of the 1928 Nobel Prize for physics or medicine, French bacteriologist Charles J. H. Nicolle (1866–1936) was recognized by the Swedish Academy for his discovery that typhus is transmitted by the human body louse and therefore could be easily prevented. Nicolle had been deaf since age 18.

7. DOROTHY CROWFOOT HODGKIN

Dorothy Crowfoot Hodgkin (1910–1994) won the Nobel Prize in chemistry in 1964 for solving the structure of vitamin B12, which cured pernicious anemia. Using X-ray crystallography, she deciphered the atomic structure of penicillin during World War II as well as the structure of insulin. The crystallographer transformed her art from spurious sideline to a powerful scientific tool. Hodgkin accomplished all this despite suffering from rheumatoid arthritis that crippled her hands and feet.

8. EDWARD TELLER

Born in Budapest, Edward Teller (1908–) attended the University of Munich to study physics. When he was a student in the summer of 1918, his right foot was severed in a streetcar accident. The amputation was so sudden that he didn't realize what had happened until he saw his boot with his foot still in it, lying in front of him. What was left of Teller's leg was reconstructed so that he could walk without a prosthesis, although he usually chose to use an artificial foot.

9. MELVIN CALVIN

American chemist Melvin Calvin (1911–1997), who was deaf, won the 1961 Nobel Prize in chemistry for his explanation of how carbon dioxide is incorporated into green plants. He also researched the possibilities of developing oil-bearing plants as alternative energy sources. Calvin's high school science teacher told him that he would never become a scientist because he was too impulsive.

10. STEPHEN HAWKING

Born on January 8, 1942, the 300th anniversary of the death of Galileo, cosmologist Stephen Hawking has attracted much media attention for his book *A Brief History of Time* and also for his frailty. As a student at Cambridge, he discovered that he had amyotrophic lateral sclerosis, which confined him to a wheelchair. His speech became nearly incomprehensible until he received a computerized voice synthesizer.

Unethical Experiments

Scientific experimentation promotes life. However, we also have used the tools of knowledge to promote selfish gain and even death.

1. GYPSIES

Nicknamed the "Angel of Death," Dr. Josef Mengele (1911–1979) remains the symbol of Adolf Hitler's Final Solution. During his 21 months at Auschwitz, Mengele dramatized his murderous policies by pouring attention on twins, who served as perfect guinea pigs for his so-called genetic experiments. Gypsy twins were immediate targets because the wanderers were thought to be as "unclean" as Jews. In one experiment, Mengele's assistant rounded up 14 pairs of gypsy twins. The "doctor" supervised how they were placed on a marble dissection table and put to sleep. Then his assistant injected chloroform into their hearts, killing them instantly. Mengele dissected the bodies to note how each organ was affected.

2. TWINS

On January 27, 1945, Soviet armies liberated survivors of Auschwitz in Poland. Among the survivors were 157 twin children, all that remained of approximately 3,000 twins who had entered the camp since 1943. At the concentration camp, twins and those with genetic abnormalities were spared the gas chambers and placed in special barracks. They received this treatment as Josef Mengele's guinea pigs for such experiments as twin-to-twin blood transfusions, injection of the eyeballs with dyes, organ rearrangement, and sterilization. When one twin died, the other would be killed and dissected.

3. PLUTONIUM TESTS

U.S. government agencies, as part of the Manhattan Project and Atomic Energy Commission, conducted radiation experiments on nearly 1,000 people from the 1940s to the 1970s. The purpose of the experiments was to gather data for military purposes concerning the biological effects of plutonium and radioisotopes. Plutonium has no medical use. Subjects were elderly prisoners, the terminally ill, or the mentally incapacitated.

During the Cold War, the military wanted to know how much radiation a soldier could endure before becoming disabled. The Pentagon turned to the University of Cincinnati, where radiologist Eugene Saenger exposed 88 cancer patients, ages 9 to 84, to massive doses of radiation and recorded their responses. The Cincinnati study probably resulted in more deaths than any other government-sponsored experiments. Medical researchers from the 1930s through the 1950s had already determined that whole-body radiation, as used in Saenger's studies, was not effective in treating most cancers.

4. PLAGUE EXPERIMENTS

During World War II, Japan's Unit 731 operated a human experimentation facility at Pingfan, China. Russian and American prisoners were infected with bubonic plague so that doctors could study its effects. Often, plague-infected organs would be extracted without an anesthetic. At least 3,000 prisoners died and were incinerated. After the war, infected rats raised at the facility were released and killed thousands of local Chinese in an epidemic.

5. **LSD TESTS**

Approximately 6,700 human subjects were used by the U.S. government in experiments involving LSD, morphine, Demerol, mescaline, atropine, and psilocybin. The tests were advertised as helping to develop and test measures against chemical weapons. Fort Detrick in Maryland had been a scientific center since 1943 out of concern that German bombs could be outfitted with biological agents. Its series of hallucinogenic drug experiments from 1953 through 1971 was second in length only to the Tuskegee syphilis study in American research scandals. Subjects were not informed when they were given a psychotropic drug and were not monitored after the tests. Several veterans attempted to sue, but they were barred from legal action because they had incurred injuries "incident to service."

6. **BORON-10**

Dr. William Sweet of the Massachusetts General Hospital insisted that he never placed research ahead of patient interests. He began experiments in 1961 using boron-10, which was thought to concentrate in brain tumors and destroy them if bombarded with slow neurons from a nuclear reactor. When Sweet injected 61 patients with boron-10, some patients retched, some emptied their bowels, and one suffered temporary respiratory failure, but all recovered. Nineteen patients endured the next stage—radiation. They were taken to the nuclear reactor in Brookhaven, where they lay in a coffin like excavation in a concrete floor, their heads over a hole that exposed them to the reactor below. None of the patients improved from this round of treatment. A new reactor was built in

1959 with an operating room. By 1961, 16 patients were treated, their skulls exposed through a surgical flap so that the radiation would penetrate. One patient, George Heinrich, died after receiving a massive dose. Sweet disavowed any ethical or legal violations, saying either his patients or their families verbally consented to the treatments. Massachusetts General Hospital also saw no wrongdoing in Sweet's treatments.

7. TUSKEGEE EXPERIMENTS

In 1997, President Bill Clinton apologized at the White House for the suffering endured by 400 African-American men who were infected with syphilis. In tests conducted by the U.S. government from 1932 to the 1970s, the infected men were deliberately left untreated so that doctors could study the disease. Some went insane. Children were infected. Attorney Fred Gary worked for decades to win a class-action suit.

8. PROPERTY THEFT

In the 1980s, businessman John Moore was treated by a Los Angeles specialist for hairy-cell leukemia. Unbeknownst to Moore, Moore's doctor had discovered a natural compound in his spleen that appeared to have great therapeutic potential. The doctor took out a patent on the cells and sold commercial rights to a biotechnology company for millions of dollars. Moore sued for property theft. In 1990, California's Supreme Court ruled five to four that Moore did not have property interest in his body parts. The stage was set for the biotechnology industry to claim a crucial right of access to human tissue.

9. LEAD PAINT STUDY

In the mid-1990s, a Kennedy Krieger study encouraged landlords to rent lead-contaminated homes to 1,907 impoverished Baltimore families with young children. The study, overseen by Johns Hopkins University, focused on West and East Baltimore neighborhoods, where researchers estimated that 95% of row houses built before World War II were contaminated with lead paint. The purpose of the study was to determine the minimum amount of lead cleanup that could be undertaken and still protect the health of children. The houses were split into four groups of row houses, each receiving varying degrees of lead cleanup. Grants were given to landlords who rented homes to families with small children. Subjects were told that the houses had been cleaned and that lead had been removed. They signed consent forms stating only that "lead poisoning in children is a problem in Baltimore"; the forms made no mention that researchers expected children in the study to accumulate lead in the blood. The subjects claimed that their children suffered learning disabilities after living there.

10. **THE SCIENCE CLUB**

In 1997, the Quaker Oats Company and the Massachusetts Institute of Technology announced a $1.85 million settlement for approximately 30 alumni of the "Science Club" at Fernald School for the mentally retarded in Waltham, Massachusetts. From 1946 to 1973, subjects were fed cereal containing radioactive tracers supplied by the Atomic Energy Commission.

Blackballed

Galileo isn't the only scientist to have been punished or ostracized for seeing "other possibilities" that often prove to be true, or at least worthy of consideration.

1. GALILEO GALILEI

Galileo (1564–1642) escaped the final cut of the Inquisition, but the trial left him a broken man. Additionally, his intensive observations of the Sun damaged his retinas, contributing to the blindness that afflicted him the last four years of his life.

2. JOSEPH PRIESTLEY

English chemist Joseph Priestley (1733–1804) demonstrated that plants absorb carbon dioxide and emit oxygen, but he retained his belief in phlogiston—that fire is a material substance—until his death. Always outspoken, his support of the French Revolution led to accusations of sedition, especially after he was elected a member of the French legislative assembly (which he didn't accept). On the evening of July 14, 1791, Bastille Day, a mob set fire to his house in Birmingham. He fled to London but couldn't find safe lodging there. In 1793, he decided to move to Pennsylvania, where he lived for the rest of his life.

3. LISE MEITNER

Nuclear physicist Lise Meitner (1878–1968) overcame a backward Austrian education aimed at producing housewives, a chauvinistic academic standard that only males could study chemistry and physics, and Nazi Germany's purge of Jewish physicists by fleeing Germany to Denmark and Sweden. At the age of 60, she deciphered the secret behind nuclear fission, but her German partner, Otto Hahn, received the Nobel Prize in 1946 for the team's work. Hahn did not cite Meitner's contribution, stressing that he alone had solved the problem.

While working together in the Berlin laboratories of the Kaiser-Wilhelm Society (later called the Max Planck Institute), Meitner nicknamed Hahn "Haehnchen," or "chicken," because Hahn couldn't

Austrian-born physicist Lise Meitner discovered the secret behind nuclear fission but her German partner, Otto Hahn, received the Nobel Prize in 1946 for the team's work.

marry her (she was Jewish) and retain his standing in the German scientific community. Although Hahn married another woman, he and Meitner still kept in touch even after he fled Nazi Germany.

In 1992, years after Meitner's death, German physicists named element 109, the heaviest known element known at that time, meitnerium in her honor.

4. NIKOLAY VAVILOV

Biologist Nikolay Ivanovich Vavilov (1887–1943) came from a wealthy merchant family and dressed in the starched collar of the old Russian professoriate, which was duly noted during the Soviet regime after the Revolution. Vavilov promoted classic genetics, while another agronomist, T. D. Lysenko (1898–1976), promoted the inheritance of acquired characteristics. "Lysenkoism" became popular because the Stalinist regime wanted to believe that people can inherit improvements acquired from their social environment, in order to spur revolutionary changes. Vavilov tried to compromise with Lysenko, but then he decided to point out errors in his rival's views. A cultural purge began in 1937 to eliminate dissenters, among them many classic geneticists. On August 6, 1940, while he was leading an expedition in the Ukraine, Vavilov was arrested. The following July, he was sentenced to death by firing squad for espionage. In 1942, his sentence was reduced to 20 years in prison. Vavilov died there of malnutrition. After Vavilov's ideas were deemed "acceptable" in the 1960s, investigators admitted that the charges had been false.

5. JAMES CHADWICK

Before he won the Nobel Prize in physics in 1935 for the discovery of the neutron, English physicist Sir James Chadwick (1891–1974) spent the duration of World War I interned at a race track in Germany, where he shared with five other men a stable intended for two horses. To maintain his morale, he participated in a scientific society formed by a group of internees. He also persuaded one of his captors to obtain a Bunsen burner for him and enlisted a fellow captive to blow air through the tube into the burner as he conducted experiments.

6. ADOLF BUTENANDT

German biochemist Adolf Butenandt (1903–1995) isolated estrone in 1928 and androsterone in 1931. When he was awarded the Nobel Prize, Nazi authorities refused to grant him permission to go to Stockholm. The reason was that the 1935 Nobel Peace Prize winner, Carl von Ossietsky, was then in a German concentration camp. After the war, Butenandt received the medal but not the monetary award.

7. J. ROBERT OPPENHEIMER

The first director of the Los Alamos National Laboratories, J. Robert Oppenheimer (1904–1967) had sympathized with communism during the 1930s. As a result, his loyalty was questioned during the McCarthy era. He lost his security clearance and could no longer work on secret projects. Nevertheless, Oppenheimer served as director of the Institute for Advanced Study in Princeton from 1947, remaining there as a professor until his death from cancer in 1967. The Enrico Fermi Prize awarded to him in 1963 was a belated attempt to atone for the "witch-hunt."

8. EDWARD TELLER

After the successful explosion of the hydrogen bomb, nuclear physicist Edward Teller (b. 1905) was ostracized by the scientific community for testifying against J. Robert Oppenheimer, whose security clearance had been revoked during the heyday of McCarthyism.

Although he led the effort to develop the atomic bomb during World War II, J. Robert Oppenheimer's membership in the Communist Party in the 1930s and opposition to the hydrogen bomb resulted in his being blackballed during the McCarthy era. He lost his security clearance and was forbidden to work on secret projects.

9. HALTON ARP

Halton Arp (b. 1924), the astronomer who catalogued peculiar galaxies, believed he had evidence of direct ties between nearby galaxies and quasars. This view opposed one of the pillars of modern cosmology—that the red shift rule applies universally to all extragalactic objects. Arp argued that quasars had been ejected from their own galaxies and that their high velocity distorted red shift readings. Not only did other astronomers dismiss his views, they also claimed that his evidence of associations between objects of different red shifts came from photographs produced by Arp himself. In 1983, Arp was barred from telescopes at Mounts Wilson and

Palomar. He left Cal Tech for a position at the Max Planck Institute in Munich. No responsible scientist believed justice had been served.

10. **BARBARA McCLINTOCK**

Barbara McClintock (1902–1992) received the Nobel Prize in medicine in 1983 for her work in mobile genetic elements. However, in 1936, when the chairman of the University of Missouri's botany department saw an announcement of the engagement of a Miss Barbara McClintock, he summoned his assistant professor into his office. He threatened, "If you get married, you'll be fired." It turned out that another Barbara was getting married. Eventually, though, facing continued discrimination, she marched into the dean's office and asked if she would ever become tenured. No, he said, you'll probably be fired when your mentor leaves. Even though she had already won an honorary doctorate, McClintock packed her bags and left.

ASTRONOMY

Noted Early Astronomers

Not all the astronomers listed below had telescopes, but they carefully studied the skies and set forth credible theories concerning planetary and stellar movements.

1. HIPPARCHUS

Greek astronomer Hipparchus (Fl. 146–127 B.C.) correctly measured the length of the year to within six minutes and logically estimated the distances to the Sun and Moon. The author of *The Commentary on the Phenomena of Aratos and Eudoxus* is also credited with a star catalog.

2. PTOLEMY

Greek astronomer Ptolemy (second century A.D.), who worked in or near Alexandria, claimed that the universe was a sphere and that the celestial bodies revolved around Earth, the center of the universe. The fixed stars uniformly circled Earth in 24 hours, and the planets moved in more complex, variable ways. His major treatise, *Almagest,* set forth the geocentric theory that went unchallenged for almost 1,500 years.

3. COPERNICUS

In 1543, shortly before his death, Polish humanist intellectual Nicolaus Copernicus (1473–1543) published *De Revolutionibus Orbium Coelestium*, explaining his theory that Earth and other planets moved around the Sun. He made his celestial observations from a turret situated on a protective wall around the cathedral of Fraudenburg, where he had been appointed a canon. Copernicus puzzled over the complex spheres that he had built to explain celestial movements because he thought that nature must be simple. He decided to examine the question from a different perspective—making the Sun the center of the planetary system. This was not a proposal that could be made lightly in an age of religious upheaval. Copernicus delayed publication of his book out of fear of the church's outcry and also because of his perfectionism. It is said that on his deathbed a copy of his book was put into his hands.

4. GALILEO GALILEI

In 1604, Galileo Galilei (1564–1642) built his own telescope, based on a Flemish invention, to examine the sky (it had been used previously to examine terrestrial phenomena to assist navigation). He saw Jupiter's moons, the moon's craters, and Venusian phases. In 1633, Galileo was put on trial by the Inquisition and, to avoid torture, declared that Earth does not move around the Sun, even though he believed otherwise. Along with Tycho Brahe (1546–1601) and Johannes Kepler (1571–1630), he showed that the heavens are not perfect and unchanging.

5. TYCHO BRAHE

In 1572, Danish astronomer Tycho Brahe (1546–1601) observed a new star and called it a nova. He estimated its distance and found that it was beyond the Moon and the planets, in a faraway region of the stars. This presented a major problem because it was believed that the heavens were unchanging and nothing new could appear. In 1577, Brahe also observed a comet and showed that it was beyond the Moon's orbit (comets were believed to be fiery exhalations that were closer to Earth than the Moon).

Brahe sought to combine the Copernican system with the Ptolemaic system. The planets were considered as revolving around the Sun, which, in turn, revolved around fixed Earth located at the center. His observatory, Uraniborg, was built on an island between Sweden and Denmark.

After his patron Frederick II died in 1588, Tycho found himself confronting the full resentment of the nobility for freely dispensing medicine to the poor. He accepted an invitation by the Emperor Rudolph II to reside at Benetek, a few miles from Prague, where a new observatory was to be built, but he died shortly afterward.

6. JOHANNES KEPLER

In 1609, Johannes Kepler (1571–1630) published *Astronomia Nova*. He claimed that planets move around the Sun in elliptical orbits at varying speeds. That meant that mathematics of static patterns would no longer suffice, paving the way for Newton's fluxions and Leibniz's differential calculus. Kepler also showed that the time it takes a planet to make one orbit increases with its average distance from the Sun in an exact way. Between the six planetary spheres, Kepler believed there were five regular geometric solids, which he learned from Plato's *Timaeus:* between Saturn and Jupiter he fit a cube; between Jupiter and Mars, a tetrahedron; between Mars and Earth, a dodecahedron; between Earth and Venus, an icosahedron; and between Venus and Mercury, an octohedron.

The German astronomer and mathematician had attracted the attention of Tycho Brahe with his dissertations on celestial orbits. In 1599, religious persecutions encouraged Kepler to accept Tycho's invitation to Prague to assist in the preparation of astronomical tables, called the Rudolphine tables. Tycho died in 1601, and Kepler continued the work alone, being appointed imperial mathematician and astronomer. Kepler possessed Tycho's papers and observations, which enabled him to establish his three laws of planetary motion.

7. SIR ISAAC NEWTON

Sir Isaac Newton (1642–1727) invented the first reflecting telescope in 1668. Newton's *Principia,* published in 1687, presented a mathematical explanation of the structure and mechanics of the universe. His law of universal gravitation made it possible to explain the structure of the solar system.

8. EDMUND HALLEY

In 1705, Edmund Halley (1656–1742) calculated that a bright comet he had observed in 1682 had also appeared in 1531 and 1607. He correctly predicted that it would return in 1758 and would return every 75.5 years from that point. He was the first to state that comets are members of the solar system, traveling in elliptical or hyperbolic orbits. Halley also greatly encouraged and financed the publication of Newton's *Principia.*

9. SIR WILLIAM HERSCHEL

Sir William Herschel (1738–1822) discovered the planet Uranus in 1781, but he first thought the planet was a comet. He was rewarded for his discovery with a large pension from King George III. William and his sister Caroline became full-time astronomers, no longer having to rely on their musical talents to earn money. In 1783, William published his first catalog of double stars, most of which had been seen before. In 1786, the first volume of his catalog of 1,000 nebulae appeared. By 1802, he had added 1,500 more.

His catalogs formed the basis of Dreyer's *New General Catalog* of 1888, which is still used today. In 1789, Herschel's 40-foot reflecting telescope was completed, and he discovered the sixth satellite of Saturn with it.

10. **PERCIVAL LOWELL**

In 1905, Percival Lowell (1855–1916) predicted that there is a planet beyond Neptune. He was at first particularly interested in studying Mars and its "canals." During the 15 years that he devoted to Martian studies, Lowell mapped the planet's surface markings in intricate detail. He observed a network of hundreds of straight lines and their intersections. He concluded that the bright areas he saw were deserts and the dark ones patches of vegetation. He said that hot water from a melting polar cap flowed through the canals toward the equatorial region to revive the vegetation. The canals, he believed, had been constructed by intelligent beings. The last years of Lowell's life were devoted to searching for a planet beyond Neptune. He analyzed the discrepancies between the observed and calculated positions of Uranus and made allowances for the perturbations of Neptune. But after examining photographs of the region, he found no planet. The search was continued by Clyde Tombaugh, who, in 1930, discovered Pluto.

Ten Planetary Pioneers

To the ancient and medieval world, the world seemed unchanging as the same planets and stars marched their way across the night sky. But when astronomers perfected strong lenses and optical instruments, they were astonished to see more than they had ever expected.

1. GALILEO GALILEI

After he built his own telescope, around 1609, Galileo Galilei (1564–1642) saw for the first time four moons orbiting Jupiter; craters, mountains, and "seas" of the Moon; and the phases of Venus. Although the planets that he spotted had been visible to the naked eye, he was the first to view them through a telescope.

2. CHRISTIAAN HUYGENS

In 1656, Dutch astronomer and physicist Christiaan Huygens (1629–1695) observed a ring around Saturn. Actually, Galileo had already seen it, but his telescope made it appear as if the ring were two "horns" on either side of the planet. Huygens enjoyed relatively little fame because he worked during the period directly after Galileo's death and just before the ascent of Isaac Newton. In the early 1650s, Huygens learned to grind telescope lenses, tutored by the Dutch philosopher Baruch Spinoza. Using his lenses, he immediately discovered a large moon circling Saturn, which he named Titan. Incorporating his lenses into extremely long telescopes (up to 23 feet long), he charted the surface features of Mars, as well as the Great Orion nebula. Huygens later modified the design of his telescope to make an "aerial" 100 feet long.

3. **SIR WILLIAM HERSCHEL**

In 1781, German-English astronomer Sir William Herschel (1738–1822) found a planet unknown to the ancient world, more or less by accident. He didn't suspect that a planet was hiding behind Saturn, and at first he thought he had found a comet without a tail. He wanted to name it Georgium Sidus, after King George III of England, but the name was opposed by non-English astronomers. However, the king appointed Herschel the royal astronomer for his discovery of Uranus.

4. **JOHANN GOTTFRIED GALLE**

In 1847, German astronomer Johann Gottfried Galle (1812–1910) won a race with English astronomers to find a new planet predicted by two theorists, Englishman John C. Adams and Frenchman Urbain LeVerrier. Neptune's existence was predictable because its gravitational force exerts a pull on Uranus.

5. **CLYDE TOMBAUGH**

After his discovery of Pluto in 1930, Clyde Tombaugh (1906–1947) was awarded a medal by Britain's Royal Astronomical Society in 1931. Only then did he begin his undergraduate education in astronomy in 1932, at the University of Kansas. He was one of the very few scientists, if not the only, to matriculate in college after making one of the most important scientific discoveries of the century.

6. GEOFFREY MARCY AND PAUL BUTLER

Geoffrey Marcy of San Francisco State University began searching for evidence of brown dwarf stars in 1989, after learning of the link between them and possible planets orbiting them. Marcy and Paul Butler of the Carnegie Institution in Washington, D.C., announced at the January 1996 meeting of the American Astronomical Society that they had found evidence of two planets that may support life. One, a heavyweight eight times the size of Jupiter, orbits the star 70 Virginis; the other, only 3.5 times the mass of Jupiter, circles 47 Ursae Majoris.

7. GEORGE GATEWOOD

French astronomer Joseph-Jéôme de Lalande (1732–1807) published a catalog of the location of 50,000 stars. Lalande 21185 seemed to be just one of those stars, except that Lalande didn't notice that its position changed nightly. However, George Gatewood, professor at the University of Pittsburgh, found not one but several planets around the star in 1996. At least two Jupiter-sized planets seemed to be moving in circular orbits around it. Not only was this the first probable planetary system outside our system, but it also appeared to be similar to ours—two gas giants on circular orbits well away from the central star.

8. MICHEL MAYOR AND DIDIER QUELOZ

Michel Mayor (b. 1942), a Swiss-born veteran of spectroscopic searches for companions to stars, and his colleague Didier Queloz observed a planet in orbit around a sunlike star called 51 Pegasi. Before it was announced in 1995, there was no proof that planets similar to our Sun's system existed elsewhere. If not for a tragic accident, Mayor would have teamed with Antoine Duquennoy, who, with Mayor, had surveyed the stars for years looking for stellar companions. However, Duquennoy died in 1994 in an automobile accident in the Alps after a day of mountaineering.

9. WILLIAM COCHRAN AND ARTIE HATZES

In 1996, after eight years of searching, William Cochran and Artie Hatzes of the University of Texas McDonald Observatory announced that they had codiscovered (along with Geoffrey Marcy and Paul Butler) the first planet on an eccentric orbit circling the solar-type star 16 Cygni B, about 70 light-years from Earth. The eccentric orbit is reportedly caused by 16 Cygni's binary companion.

10. ROBERT NOYES

In 1997, Rho Coronae Borealis was found to have a companion with a mass of at least 1.1 Jupiter masses moving in a circular orbit with a period of 40 days. Harvard-Smithsonian's Robert Noyes and his team found that the star's planet had the largest orbital radius of the hot Jupiters found orbiting other stars, pointing out that hot Jupiters did not reside at the same distance from their stars. Giant planets might be found in orbit anywhere between their star and point of origin.

Remarkable Observational Tools

Since Galileo built his telescope in 1609, astronomers have been improving and expanding the magical mirrors and lenses.

1. KEPLER'S EYEPIECE

In 1611, Johannes Kepler (1571–1630) substituted a convex eyepiece for a concave one. This enlarged the field of view but turned the image upside down.

2. NEWTON'S REFLECTOR

In 1688, Sir Isaac Newton (1642–1727) built a reflecting telescope using a concave mirror instead of a lens. The reflector eliminated a rainbow-colored rim around images that appeared in refracting telescopes. The mirror is also less prone to distort large images.

3. HERSCHEL'S '49ER

In 1789, Sir William Herschel (1738–1822) built a telescope with a 49-inch mirror. He used it to discover Encelades and Mimas, two of Saturn's moons.

4. **PARSON'S REFLECTIVE TELESCOPE**

In 1845, William Parsons, the earl of Rosse (1800–1867), built a reflecting telescope with a 72-inch mirror, the largest in the world until 1917.

5. **LICK OBSERVATORY**

In 1888, a 36-inch refracting telescope was completed at Lick Observatory in the Diablo Range east of San Jose, California. Eccentric California millionaire James Lick issued final orders on his deathbed for the dispensation of his fortune, earned mostly from real estate. His largest bequest was for the construction of the observatory that bears his name. Trained as a cabinetmaker, Lick built his fortune after his marriage proposal was turned down by the prospective bride's father. He had previously thought of erecting a monumental statue of himself or a pyramid larger than that at Giza. His astronomer friend George Davidson persuaded him to build the observatory instead.

6. **MOUNT WILSON**

In 1908, a 60-inch reflector was completed at Mount Wilson. George Ellery Hale (1868–1938) had built his first telescope in his backyard, but it didn't work well. He asked the help of his neighbor, who turned out to be S. W. Burnham (1838–1921), a famous double-star observer. Burnham told Hale's father that a four-inch refractor was available. After his father bought the telescope, Hale was hooked for life. Academics bored him at MIT, so Hale went to Harvard, where he tested his spectroheliograph, which photographs the Sun in the light of a single element. He focused on studying the Sun and its properties and sought to combine astronomy with physics. Mount Wilson had been found to be a forbidding place because it wasn't easily accessible, but Hale shaped it into a world-class observatory after signing a 99-year lease with the Mount Wilson Toll Road Company for 40 acres on the mountaintop; the landowners provided the property rent free because they thought the observatory would attract tourists to the Pasadena area. Founded in 1904, the site was first devoted to solar studies. With the addition of a 100-inch telescope in 1917, "solar" was dropped from the observatory's name.

7. **METAL DISH**

In 1936, after designing the world's first radio telescope, Illinois engineer Grote Reber erected a 30-foot metal dish in his backyard and mapped the Milky Way, a project that took him eight years. An engineer employed by a Chicago radio manufacturer, Reber had read Karl Jansky's study of radio signals. Reber not only confirmed that the galactic nucleus was the signal source, but located other "hot spots" in the sky that did not appear to coincide with any visible objects. His first reports were published in 1940 but were not followed up during the war years.

8. **THE HALE**

A 200-inch reflector was built in 1948 at Mount Palomar. For nearly a half century, the world's most powerful research scope was the Hale on Palomar Mountain in California. Its 26-inch-thick, 20-ton mirror focused on distant starlight that mere 100-inch scopes could not capture. As a result of the work of George Ellery Hale in 1928, a grant was awarded for the construction of the 200-inch telescope. In 1934, Corning Glass Works in New York developed a special casting technique for the huge Pyrex glass disk. Cooled for eight months, the 20-ton disk was shipped by rail to Pasadena to grind and polish. However, the mirror showed huge optical distortions, which made the telescope almost useless. A second mirror was ordered from Corning, cast in one piece, but because casting had to be done from several melting pots due to the size of the piece, the mirror broke during cooling. A third mirror was ordered, this time from Schott in Mainz, Germany. This was a very delicate job, because the United States was at war with Germany. An agreement between Roosevelt and Hitler made the job possible.

The secret service hid the transport from Mainz to Hamburg. The mirror, yet uncoated, was shipped to San Diego with a warship escort early in 1945 in the last days of the war. A special coating process deposited aluminum uniformly so as not to change the surface's shape. It was not until 1947 that the mirror was installed. Only a small engraved plate inside the observatory tells the real story.

9. HUBBLE SPACE TELESCOPE

In 1990, the Hubble Space Telescope was launched from the space shuttle *Discovery*. The orbiting observatory's 8-foot reflector produced ultrabright images because of no atmospheric obstruction. The telescope was almost useless at first because it was damaged during launch. It was repaired in open space—an astronaut maneuvered it onto the shuttle's loading dock, added an aspherical correction lens, and reejected it into space.

10. KECK TELESCOPES

At the summit of Hawaii's dormant Mauna Kea volcano sits the W. M. Keck Observatory, which holds the world's largest optical and infrared telescope. The primary mirror measures 10 meters in diameter and comprises 36 hexagonal segments that work in concert as a single piece of reflective glass. This process, called "adaptive optics," reduces blurring and atmospheric distortions. Operated by the California Institute of Technology, the University of California, and NASA, Keck 1 began observation in May 1993 and Keck II in October 1996. Astronomers use the telescopes in shifts of one to four nights. The twin Keck telescopes each measure 33 feet in diameter. In 1999, Geoffrey Marcy and Paul Butler predicted and observed the transit of a massive exoplanet (or planet outside the solar system) using the Keck telescope.

Early Observational Instruments

Many observational tools other than telescopes have been used to chart the heavens. Evidence is strong that the hidden north passage in the Great Pyramid at Giza is aligned with the lower culmination of the Pole Star at the time the pyramid was built. The alignment of the pyramidal bases on the cardinal directions was probably laid down by careful astronomical observations by the Egyptians.

1. STONEHENGE

The best-known evidence that early man used the sky as a calendar is surely Stonehenge. For more than 100 years, the secret of its alignment with the summer solstice has been known. But Stonehenge doesn't stand alone. There are at least 400 other structures like it throughout the British Isles. Bronze Age astronomy was fully as advanced in the British Isles as in the Fertile Crescent.

2. BIGHORN MEDICINE WHEEL

Almost every visitor to the Bighorn Medicine Wheel in Bighorn National Forest is disappointed because at first glance it seems to be a crude structure that could have been be built by almost anyone. Closer observation reveals that its builders had a specific purpose in mind. The wheel includes a hub, a pile of rocks about 12 feet across, hollowed out in the middle. From it radiate 28 spokes that end in a ring approximately 90 feet in diameter and almost elliptical in shape. Six rock cairns sit along the wheel's periphery, opening in various ways. The central cairn seems to mark the summer solstice, the time of the sun dance ceremony. Additionally, a second cairn points to where the sun would set on that day.

3. TOWER OF THE WINDS

The octagonal tower, which stands in Athens, is mentioned by Vitruvius, a Roman architect and engineer of the first century B.C., who attributes its design to Androniclus, a Macedonian astronomer. Its outer surface is decorated with five planar sundials and figures personifying the winds. The placement of holes and grooves suggests that the tower once housed an elaborate water clock. A semi-cylindrical structure contained a water tank and a series of floats, pulleys, and weights generating a force that drove a clock. When the clock operated, the flowing water and pulleys caused the star map, with its sun symbol, to rotate once every 24 hours.

4. ARMILLARY SPHERE

From 52 B.C. to 132 A.D., Chinese astronomers devised an armillary sphere for measuring the position of heavenly objects. At its center was a metal ring representing the equator, and rings representing the planets' paths radiated from the center. The devise also included a water clock.

5. THE TRIQUETRUM

In 150 A.D., Ptolemy marked the stars' positions with a triangular rule, called a triquetrum, and a plinth, a block of stone inscribed with a calibrated arc, to plot the elevation of the Sun.

6. ASTROLABE

In 927, Nastulus, a Muslim instrument maker, fashioned the oldest known astrolabe, a metal map of the heavens showing the stars' apparent movement around the Pole Star and the positions relative to the horizon.

7. URANIBORG

In 1576, Tycho Brahe (1546–1601) began construction of Uraniborg, an observatory on the island of Hven, given to him by Denmark's Frederick II. His equipment included a wall quadrant, an armillary sphere, and a sextant covering 30 degrees of the sky. The island also included a smaller observatory called Stjerneborg, the castle of the star. But Tycho's abrasive personality cost him; after Frederick's death, he was forced off the island. Tycho received funds from Polish emperor Rudolph II to build another observatory near Prague, but he was unable to complete construction before his death.

8. DE DONDI'S ASTRARIUM

In 1362, Paduan clockmaker Giovanni de Dondi (1318–1389) completed 16 years of work when he unveiled his astronomical clock that not only told time but also recorded the movements of the planets. Fifty inches high, the clock's dials traced the movements of Mars, Venus, and the Sun. The clock moved a wheel (making one turn a day) that moved another wheel that moved pointers representing the stars' positions.

9. JAIPUR AND DELHI OBSERVATORIES

The Jaipur Observatory, known as Yantra, is the largest of five built by Jai Singh II, Maharaja of Jaipur. Erected in 1734, the structures provided fixed angles to check the stars' positions. The Delhi Observatory, built by Singh in 1724, is a 56-foot-high gnomon, or sundial. It casts a shadow on masonry arcs calibrated in hours, minutes, and seconds. The sunken bowl at its foot is marked with circles of celestial longitude and latitude. A perforated metal disk held over the bowl casts a pinpoint of light inside, indicating the sun's position. Singh hoped to improve the Indian calendar by obtaining accurate measurements of celestial motions, but he did not finish the project.

10. **GALILEO'S TELESCOPE**

Having heard about Dutch optician Hans Lippershey's (c. 1570–c. 1619) large optical lens (invented in 1608), Galileo Galilei (1564–1642) devised his own telescope. When he demonstrated the device to the Doge and the Senate of Venice, they immediately saw its commercial and military uses. Unlike Lippershey, who was denied a license to manufacture the instrument, Galileo doubled his salary and was granted tenure at the University of Padua.

Models of the Universe

The stars appear to be at the same height if you look up at the sky from the ground. Adding a third dimension, depth, led astronomers to speculate about the structure of the universe. Our own galaxy, the Milky Way, was widely agreed to constitute the entire universe up to the early 1900s. The term *universe* meant something quite different in the ancient and medieval world than what it does now.

1. **HIERARCHICAL**

At the beginning of the twentieth century, Swedish astronomer Charles Charlier (1862–1934) proposed that the universe is grouped in successively larger clumps (similar to the way Russian dolls nest inside each other). Stars group together, he said, to form galaxies, galaxies group in clusters, clusters of galaxies form superclusters, and so on. If this were true, the universe would be infinite. There is evidence of clustering up to the scale of superclusters, and the smoothness of background radiation could suggest that the superclusters are distributed evenly across space.

2. PANCAKE

By 1919, American astronomer Harlow Shapley (1885–1972) had calculated the distance to many stars in our galaxy and announced that he had found the shape of the universe. He theorized that just as planets were grouped around the center of the solar system, globulars (globe-shaped star clusters) were grouped around the center of the universe.

The sun wasn't the center of the universe, as previously thought. Placing stars in relation to each other, Shapley built a three-dimensional model of the universe. Viewed edge on, it resembled a pancake with a bulge at the center. What Shapley called the universe was actually our galaxy. He estimated it to be 250,000 light-years across, with the Sun being 50,000 light-years from the center. Later calculations reveal a diameter of 100,000 light-years and the Sun's distance at 30,000 light-years from the center.

3. OPEN, CLOSED, AND FLAT

In 1922, Russian theorist Alexander Friedmann (1888–1925) derived his models of the universe from Einstein's equations of general relativity without the cosmological constant. Because he excluded this stabilizer, Friedmann described a universe in constant motion. His open model begins as a point, at zero, then expands forever. By contrast, the closed model bursts outward from a speck, expanding until growth comes to a halt, then shrinks to a point. The flat model starts at zero, then, although it continues to expand, teeters forever on the brink of collapsing.

4. EXPANSION

In 1924, Edwin Powell Hubble (1889–1953) used the Cepheid variable method to determine how far away other galaxies are from the Milky Way. His technique involves choosing a Cepheid star (variable stars), measuring intervals between the star's light bursts, and then determining the star's brightness and rate of expansion. In 1929, Hubble combined his galactic distance data with light spectra information gathered by Vesto Melvin Slipher (1875–1969), who discovered that the more distant a galaxy is, the greater its radiation is shifted toward the red end of the spectrum. Thus, galaxies are speeding away from us. Because Hubble believed that the Milky Way occupies no special place in the cosmos, all galaxies must be moving away from each other, as well as the Milky Way. The rate of expansion is named the Hubble constant, a universal number in cosmology.

5. **STEADY-STATE**

In 1949, British astronomers Fred Hoyle, Thomas Gold, and Hermann Bondi proposed the steady-state model as an alternative to theories of expansion. To them, a big bang meant that matter and energy were instantly created from absolutely nothing. They suggested a universe that is basically the same for all times. As galaxies recede, new matter is created to fill in the voids.

6. **STANDARD**

Generally accepted internationally since the 1960s, the standard model of cosmology postulated a universe that originated in a hot big bang. Essentially, it is an extension of the big bang theory. Prominent contributions to the model include those of Yakov Zel'dovich, Steven Weinberg, Joseph Suk, and Denis Sciama. The standard model traces the history of the universe toward an incredibly small "Planck time" (10^{-40} seconds after the big bang) in which the usual laws of physics supposedly break down.

7. **INFLATION**

The inflation model was developed in the 1980s by theorists Alan Guth of the Massachusetts Institute of Technology and Andrei Linde of Moscow University, as well as Paul Steinhardt and Andreas Albrecht of the University of Pennsylvania (the first two working independently). They theorized that the cosmos underwent a period of rapid expansion during the initial seconds after the big bang. Then, it expanded at a much slower rate. This forces the universe to become much flatter. After all, cosmic background radiation appears approximately the same no matter where it is observed.

8. MONOPOLE

In the mid-1990s, Russian-born cosmologist Andrei Linde (b. 1948) developed the idea of the universe as a self-reproducing system that sprouts other inflationary universes and is itself a sprout from another universe. This model, called the monopole universe, states that the entire universe exists inside a single magnetic monopole produced by inflation. Such a monopole would resemble a magnetically charged black hole, connecting our universe through a wormhole in space-time to another region of inflation.

9. SUPERSTRINGS

String theory combines general relativity and quantum mechanics into a unified theory of fundamental forces. An important feature of string models is that they predict 10 to 11 dimensions. Four of those are the three spatial dimensions, plus time. The remaining six or seven are hidden. In some theories, they are rolled up like a ball whose radius is too tiny to be detected.

These lines of concentrated energy, or cosmic strings, could have been formed shortly after the big bang and might still be present. They are thin tubes of symmetrical, high-energy vacuums without ends. They either form closed loops or extend to infinity.

T. W. B. Kibble of the Imperial College of Science and Technology in London pioneered work on the string theory in a 1976 paper. No cosmic strings have yet been detected, but they could be detected by their bizarre gravitational properties.

10. **THE M-THEORY**

The M-theory adds an eleventh dimension to the string model. Ordinary matter is confined to two three-dimensional surfaces, known as "branes" (membranes), separated by a small gap along the eleventh dimension.

Astronomical Surprises

It's human nature to worship the regularity we observe in the planets' motions around the Sun, but even in this clocklike realm, scientists have detected chaos. Pluto confounded our expectations of eternal clockwork. Over time—that is, a billion years—the planet's orbit could end up nowhere near where it was predicted. Scientists began looking for reasons behind this random path and realized that chaos was at work. Even the tiniest change in initial conditions—such as an asteroid's orbital period in relation to Jupiter's—can lead to different long-term results. Here are other mysteries of the universe.

1. HOMUNCULUS

During the 1830s, Sir John Herschel (1792–1871), Sir William's son, noted that Eta Carinae, an inhabitant of the southern skies, seemed to be surrounded by a nebulous gas cloud. Because of its resemblance to a human form, later observers called the nebula the Homunculus, or little man. Over time, Homunculus has changed in shape and brightness. The density of the core cloud and oddities in the region's spectrum have fueled speculation by some that Eta Carinae may not be alone. In fact, it may be a stellar nursery, where violent births emit clouds of obscuring dust and gas. Others believe that Eta Carinae is an older star ejecting its atmospheres or a middle-aged star stealing matter from an unseen companion.

2. PLANET X

Astronomers have proposed the existence of a Planet X, the so-called tenth planet. The unseen orb explains anomalies in the behavior of Uranus, whose calculated orbit works for one circuit but not the next, and for Neptune, whose positions are noticeably adrift after several years. Planet X may be found, scientists say, in the southern sky near the constellation Centaurus.

3. NEMESIS

Far beyond Planet X resides a dangerous killer, Nemesis, or the Death Star. This dark stellar companion to the Sun is theorized to return to the solar system every 26 million years, passing on its way through the Oort cloud, a huge sphere of cometary material, and sending a rain of comets toward the planets. The only evidence of Nemesis is the mass-extinction pattern of 26 million years. Some theorize that Nemesis could be a brown dwarf that emits little or no visible light.

4. BLACK HOLES

When the energy supply of a massive, aging star declines, it no longer produces sufficient heat and outward pressure to counteract the gravitational pull of its own particles. The star collapses, drawing inward as its gravitational field builds. Finally, crushed to tremendous density, the star is extinguished. In its place is the space-time abyss known as a black hole. At the black hole's center is a vanishing point whose density is so enormous that it causes space-time to turn in on itself.

5. INTERACTING GALAXIES

In the 1950s, Fritz Zwicky, a Bulgarian-born astronomer from the California Institute of Technology, catalogued oddball galaxies, many of which seemed to interact with close companions. Contrary to popular belief, Zwicky proposed that galaxies did not live in isolation but slipped and slid into each other, sometimes changing the course of each other's evolution. Key to his theory were bridges, or links, that Zwicky observed between many galactic companions. Because Zwicky was not popular and was prone to eccentric behavior—he had ordered an assistant to fire a shot from a rifle along the line of sight of the Hale telescope to improve the view—no one was eager to support his claims.

6. CLOSE ENCOUNTERS

Halton Arp, another Cal Tech researcher with unconventional views, gathered a collection of weird galactic types from Mount Palomar's lenses. Arp's 1966 photo album, titled *Atlas of Peculiar Galaxies,* documented warped structures, which he colorfully named "The Antennae," "The Playing Mice," "The Telephone," "The Carafe," and "The Fly's Wing." Arp also found more than 300 interacting galaxies, pointing to close encounters of the cloud kind. Arp said that the Milky Way itself seems to be involved in a prolonged affair with southern companions, the large and small Magellanic Clouds.

7. DARK MATTER

There's more to the universe than meets the eye. Until the 1980s, it was widely accepted that most of the universe could be studied by its emission of light or other forms of electromagnetic radiation. But now it's clear that less than half of the universe may be matter as we know it. Some theorists say that two-thirds or more of the universe's mass is in the form of particles that have never been detected. They go by names such as axions and gravitrons. Instead of matter, it is light that is missing from the picture. Dark matter may arise if a brown dwarf fades into oblivion or into a black hole.

8. THE GREAT WALL

Located in the opposite part of the sky from Perseus-Pisces-Pegasus, the so-called Great Wall is among the largest coherent structures yet discovered. Together with the Pisces-Perseus chain, the structures seem to form a giant arc that encircles Earth. Arrayed across the cosmos in the form of a vast, crumpled membrane, the system of thousands of galaxies defies theoretical efforts to explain its great size. Astronomers believe that the wall may be too massive to have been formed by mutual gravitational attraction of member galaxies. Others surmise that the structure consists of looser groups of galaxies than what appears today.

9. EINSTEIN'S RINGS

Rings or arcs around clusters of galaxies arise from an effect known as gravitational lensing, which occurs when gravity from a massive object binds light passing by. When a cluster of galaxies blocks our view of another galaxy behind it, the cluster's gravity warps the more distant galaxy's light, creating rings or arcs. The nearer cluster acts as a telescope, bending light into our detectors.

10. WORMHOLES

A shortcut through space-time, a wormhole is a cosmic portal connecting two black holes, or one black hole and a white hole. The "other end" of a wormhole could be anywhere in space or time, allowing an object that passes through the portal to appear instantaneously in another part of the universe.

Exoplanets

Astronomers have long suspected that other stars are orbited by planets, but firm evidence eluded us until a few years ago. Planetary systems are so common that Earthlike planets may number in the millions. Just as the late Carl Sagan said, there may be billions and billions of stars orbited by billions of planets. Listed below are stars around which planets have been detected.

1. HD 168443

HD 168443 is a solar-type star 123 light-years away in the constellation Serpens. The HD 168443 planetary system consists of two heavyweights. One planet, found in 1998, weighs between 7.7 and 15 Jupiter masses, while the other, found in 2001, which orbits farther from the star, weighs 17 to 40 Jupiter masses.

2. HD 209458

In 1999, planet hunters Geoff Marcy of the University of California at Berkeley; Greg Henry of Tennessee State University; Paul Butler of the Carnegie Institution of Washington, D.C.; and Steven Vogt of the University of California at Santa Cruz found an object orbiting a star called HD 209458 in the constellation Pegasus. When the planet passed in front of its star, it cast a shadow that was visible to observers on Earth.

3. GLIESE 876

The red dwarf Gliese 876 is the closest star to Earth, located in the constellation Aquarius, 15 light-years away and 20 million miles from the Sun. This star is 5,000 degrees cooler, 100 times dimmer, and one-third the mass of the Sun. The planet that revolves around Gliese 876 falls within the habitable zone. The giant orbits its sun in 61 days in an unusually elliptical path around the red-dwarf star. Berkeley astronomer Geoffrey Marcy's team discovered this planet in 1998. In 2001, the team discovered another planet that orbits its sun in 30.1 days, about half the orbit time or period as the first planet. It is believed that the two shepherd each other to maintain a synchrony.

4. HD 210277

Discovered by the Keck I telescope atop Hawaii's Mauna Kea in 1998, HD 210277 averages the same distance from its parent star as Earth's distance from the Sun. The planet is far heavier than Earth and has much more of an elongated orbit. Its parent star is 68 light-years from Earth.

5. TAU BOOTES

Located in the constellation Bootes, 50 light-years from Earth and 4.3 million miles from the Sun, the star Tau Bootes is three times as bright as our Sun. Its planet is approximately 3.8 times as massive as Jupiter, but it circles Tau Bootes every 3.3 days at a distance of 6.8 million kilometers. That's one-tenth the distance at which Mercury orbits our Sun. Another interesting note is that Tau Bootes has an abundant supply of elements heavier than helium.

6. 51 PEGASUS

Astronomers Michael Mayor and Didier Queloz of Switzerland's Geneva Observatory discovered the first planet around a sunlike star in the constellation Pegasus, 50 light-years from Earth and 5 million miles from the Sun. Its mass may be equal to 140 Earths.

7. EPSILON ERIDANI

Roughly the size of Jupiter, this planet revolves around a star that is near enough to be seen with the naked eye and is about the size of our Sun. The planet orbits 300 million miles from the star and takes about seven years to complete one circuit.

8. UPSILON ANDROMIDAE

In 1996, Geoffrey Marcy and Paul Butler placed three planets around Upsilon Andromidae, 44 light-years away in the constellation Andromeda. One planet, they speculated, weighed three-quarters of Jupiter's mass and was orbiting the star every 4.6 days. Two additional planets have been found since their discovery. One weighs twice as much as Jupiter and orbits the star every 242 days. The second weighs twice as much as the first and completes one orbit every four years. What's puzzling is that Jupiter-like planets usually circle more than 4.2 astronomical units from a sun, while this giant circles more closely.

9. HD 83443

In 2000, Geneva Observatory's astronomers announced the discovery of two planets circling HD 83443 in the constellation Vega. One is roughly the size of Saturn and orbits closer to the star, while the other weighs in at just 0.15 of the mass of Jupiter. One takes less than three days to circle the sun, the other, 30.

10. EPSILON RETICULUM

Located about 59.5 light-years from the Sun, Epsilon Reticuli lies in the northeastern corner of the constellation Reticulum, the Reticule, or Net. On December 11, 2000, astronomers announced the discovery of a large planetary companion to this star in a stable, highly circular orbit. The middle-weight planet lies in an Earth-like orbit inside the "habitable zone" where liquid water could exist. But the planet itself is not Earthlike: It weighs at least 1.26 times the mass of Jupiter and takes a leisurely 426 days to complete the voyage around its star.

PHYSICS

Building Blocks of Matter

While many of us dutifully learned that the basic building blocks of atoms are electrons, protons, and neutrons, which circle neatly around the center of atoms, scientists looked further into atomic structure to find what's really inside atoms. Subatomic particles are smaller than atoms. They include not only electrons, protons, and neutrons but also elementary particles.

1. BOSONS

Bosons, named in honor of Satyendra Nath Bose (1894–1974), are a group of particles with an integer-number spin—0, 1, 2. Contrary to bosons is the group of particles with a half-number spin—$1/2$, $3/2$—called fermions, in honor of Enrico Fermi (1901–1954).

2. LEPTONS

A lepton is any class of light elementary particles that are not affected by a strong nuclear force; they do not interact strongly with other particles or nuclei. Leptons include the electron, muon, and tau, plus their neutrinos and their six antiparticles.

3. MESONS

A meson is any group of unstable subatomic particles that are intermediate in mass between an electron and a proton. Mesons and baryons compose the hadron family, which interact strongly with each other and form the building blocks of matter. Leptons, however, contain particles that interact with each other through a weak force. They are responsible for particles' activity and for keeping particles together.

4. BARYONS

A baryon is a class of elementary particles that includes the proton; the neutron; and a large number of unstable heavier particles, known as hyperons. They are subject to a strong force and are thought to be a combination of three quarks.

5. MUONS

A muon is a lepton that appears to be almost identical to the electron, except that is unstable and has a resting mass approximately 207 times greater. It is the chief constituent of cosmic radiation at Earth's surface.

6. NEUTRINOS

A member of the lepton family, the neutrino has no electric charge or mass and interacts with other particles. Quantum physicist Wolfgang Pauli (1900–1958) suggested the need for a new particle to carry off the excess energy that occurred during beta decay. Neutrinos are extremely difficult to detect, even though they are produced copiously in nuclear fission reactions. The neutrino's probability of interacting is so small that the possibility of detecting a reaction is very poor, even if huge numbers of neutrinos are present. Their interaction is so weak that they can pass through Earth without hitting anything. There are three "flavors" of neutrinos, one each associated with the electron, muon, and tau particle.

The neutrino particle is not to be confused with a neutron, which is an elementary particle found in all atomic nuclei, except that of hydrogen. The neutron has no electric charge. Outside the nucleus, a neutron will decay into a proton, an electron, or an antineutrino. Each neutron comprises three quarks and is a member of the baryon family.

7. TAU PARTICLES

A lepton that is a heavy counterpart to the electron, the tau particle, was detected by Martin Perl, and in 1974, he called it the "U" particle for "unknown." He continued his quest to confirm the tau particle's existence, which is extremely difficult because it seldom interacts with other particles. Its existence was surprising to physicists because they had believed that there were two families of particles: ones with two quarks and two leptons, and those that were lighter (with up and down quarks) or heavier that consisted of charm and strange quarks. The discovery of the tau particle led to the search for a third set of quarks: bottom and top.

8. GRAVITONS

A graviton is a hypothetical particle required by quantum theory to carry the force of gravity between two objects that have mass. A type of boson, gravitons carry out the equivalent role for gravity that photons do for the electromagnetic force between two charged particles.

9. POSITRONS

As named by Paul Dirac (1902–1984) in 1929, a positron is a subatomic particle that was predicted and then later found in 1932.

The particle is an antimatter mirror image of an electron with the same mass, yet opposite charge.

10. QUARKS

A quark is a fermion with a spin $1/2$ or $3/2$. Quarks come in six varieties, which are related to one another in three pairs: (1) "up" and "down," the counterparts in the quark family of the electron and its neutrino in the lepton family; (2) "strange" and "charm," with lepton counterparts the muon and its neutrino; and (3) "top" and "bottom," with lepton counterparts the tau particle and its neutrino. Together, these combine in threes to compose other hadrons. Quarks also have "color," analogous to electric charge. It is currently deemed impossible for an isolated quark to exist.

Atomic Theories

Atomic theory has had fits and starts in order to achieve the position it is in today. Often theories are quickly discarded for better ones, but students can gain knowledge from previous attempts at explanation.

1. HOOKS THROUGH A VOID

Democritus (c. 460–c. 370 B.C.) developed the ideas of his master, Leucippus, that atoms are what everything comprises, even the human soul. Each ever-moving particle had shapes and qualities of its own, giving it distinct properties. Water atoms were smooth and round, so they flowed over one another, while iron atoms were hooked and jagged, so they stuck together. He believed that white things had smooth atoms while sour things had needle-like atoms. He also believed that atoms could not be created or destroyed.

2. FOUR QUALITIES

Aristotle's (384–322 B.C.) ideas of the four elements said that all matter had qualities associated with each—earth, air, water, and fire. Further, four qualities described each: hot and dry were associated with fire; hot and moist, with air; cold and moist, with water;

and cold and dry, with earth. The heavens consisted of a fifth element, aether. Aristotle's theory of the formation of metals was still used in Europe in the 1600s. His theory remained persuasive well into the seventeenth century.

3. SPIRIT ATOMS

While heating wood, but not lighting it directly on fire, alchemists of the Middle Ages noticed the vapor and tar rising and developed a theory of philosophical essences and spirits. This is the first time someone described the principles of composition contrary to the terms of Aristotle's earth, air, fire, and water.

4. THREE PRINCIPLES

Paracelsus (1493–1541) devised his own theory concerning the contents of matter, saying that it sometimes consisted of fire, but always included a mixture of air, water, and/or the three principles of salt, sulfur, and mercury, known as the Tria Prima.

5. NOT SO ELEMENTARY, WATSON

Johann Joachim Becher (1635–1682), a German professor and physician, was full of ambition and imagination, but bad disposition and unpractical character gave him enemies and chased away patrons. He said that all earthy substances are compounds, and there are no single elements. All mineral substances are composed ultimately of earth and water, but there are three possible earths: terra prima, stony; terra tertia, fluid; and terra secunda, fatty. Terra secunda is the essence of terra prima and imparts combustibility. Becher's theories are said to have influenced Stahl in developing the phlogistic theory, thus influencing the history of chemistry.

6. PHLOGISTON THEORY

Believed in by many after Aristotle's long-lived popularity had weakened, phlogiston, or the principle of combustion, was an invisible material, or what actually burned. It was believed to be a real substance that transferred from one chemical to another, or could be picked up from fire. It was also once believed to have negative weight.

7. CALORIC THEORY

Antoine-Laurent Lavoisier (1743–1794) postulated that heat is a weightless fluid called caloric. Each atom has a layer of caloric around it. When heated, the caloric layer gets larger, increasing the volume of the atom and decreasing the density of the substance.

8. HYDROGEN THEORY

English physician William Prout (1785–1850) pointed out that most elements had atomic weights in multiples of the atomic weight of hydrogen. He suggested that all atoms consist of hydrogen atoms, somehow packed together.

9. THE CART BEFORE THE HORSE

Dmitri Mendeleev's (1834–1907) Periodic Table, published in 1869, was not immediately accepted. (Julius Lothar Meyer [1830–1895] published similar conclusions a year later.) His results were so accurate that while building the table, he saw three places where it appeared that elements were missing. Mendeleev predicted the elements' valence, color, atomic weight, and density. In 1875, the first of the missing elements, gallium, was discovered, closely matching the Russian's predictions. Chemists finally took notice of the table. Lucky for Mendeleev, scandium (1879) and germanium (1896) were found in his lifetime, proving that he was correct.

10. **PLUM PUDDING**

As he studied the passage of electricity through a gas in a cathode ray tube, English scientist Sir Joseph John "J. J." Thomson (1856–1940) discovered that the atom has negatively charged particles. He called them "corpuscles" (now known as electrons). This was the first instance of particles residing inside the atom—most people agreed that the atom, the most fundamental unit of matter, was indivisible.

Thomson concluded that there would also have to be positively charged particles to balance out the negatively charged particles in the atom, but he could never find them. As a result, he proposed the "Plum Pudding" model of atoms, with electrons in a positively charged "sea," like plums in pudding. In 1897, Thomson determined the charge-mass relation for electrons, later to ions and canal rays, finding that the electron mass is only a fraction of the atomic mass. In those times, the discoveries were so fast in the field of gas discharges and atomic physics that it is difficult to track, and some citations are not consistent.

The Bomb Squad

When Otto Hahn, Fritz Strassman, and Lise Meitner produced and explained nuclear fission (Hahn and Strassman won the Nobel Prize in 1944), the news swept through the physics community like wildfire. Physicists realized the military potential of nuclear energy and convinced the U.S. government to see if nuclear fission could be used in the production of weapons. President Franklin D. Roosevelt appointed an advisory committee that gave a favorable recommendation. Additionally, on August 2, 1939, Albert Einstein and Leo Szilard wrote a letter to Roosevelt, explaining the possible use of energy gained from a nuclear chain reaction for a bomb, and the Manhattan Project was created to pursue the development of the world's first atomic device.

1. NIELS AND AAGE BOHR

Albert Einstein credited Niels Bohr (1885–1962) with having a "rare blend of boldness and caution, and seldom has anyone possessed such an intuitive grasp of hidden things combined with such a strong critical sense." The Dane received the Nobel Prize in physics in 1922 for the quantum mechanical model that he had developed a decade earlier, the most significant step in understanding atomic structure since John Dalton proposed the modern atomic theory in

1803. Bohr contributed to the development of the liquid-drop model of the atomic nucleus, a model used in the explanation of nuclear fission. Born to a Jewish mother and German father, Niels left Denmark in 1943, being smuggled out of the country to Sweden aboard a fishing boat and flown to England in the bomb bay of a Mosquito bomber. Bohr and his son Aage worked on the Manhattan Project after they made their way to the United States.

2. LEO SZILARD

In 1942, Hungarian-born Leo Szilard (1898–1964) set up, with Enrico Fermi, the first nuclear chain reaction. He had left Columbia University to become part of the Manhattan Project. Although he feared that further experiments would produce terrible tragedies, his pleas for a demonstration test of an atom bomb to scare the Japanese into surrender fell on deaf ears. After the war, Szilard turned to molecular biology.

3. ENRICO FERMI

Italian-American physicist Enrico Fermi (1901–1954) developed a statistical method for describing the behavior of a cloud of electrons that came to be known as the Fermi-Dirac statistics, which was also used to analyze a system of particles. Fermi built the first nuclear reactor in the basement of the stadium in downtown Chicago, where the first self-sustaining nuclear chain reaction took place on December 2, 1942. His study of nuclear changes during bombardment of nuclei brought him into the Manhattan Project and also won him the 1938 Nobel Prize for physics.

4. J. ROBERT OPPENHEIMER

Theoretical physicist J. Robert Oppenheimer (1904–1967) conducted research on protons and their relation to electrons, which led to the discovery by his students of a new particle, the positron. He also discovered that it was possible to accelerate deuterons, comprising a proton and neutron, so that they could bombard positively charged atomic nuclei at high energies.

Oppenheimer's charismatic presence and many accomplishments in the field led to his appointment as the director of the research laboratory at Los Alamos to develop the first nuclear weapon. His ability to provide a cooperative environment for secret research led to the relatively quick development of the atomic bomb. In 1945, a test device was exploded at Alamogordo, New Mexico. Oppenheimer linked its power to Shiva: "I am become death, the Shatterer of Worlds." Oppenheimer was one of a panel of four scientists who voted to use the bomb on Japan, rather than invade. After the war, he opposed proliferation of nuclear arms, a position that cost him dearly; he was accused of being a Soviet agent and relieved of his security clearance.

5. **RUDOLF PEIERLS**

German-born physicist Rudolf Peierls (1907–1995) lay on a hill in New Mexico on June 16, 1945, and watched through darkened glass as the first atomic bomb exploded at Alamogordo. In 1940, Peierls and associate Otto Frisch estimated the energy released in a nuclear chain reaction using the 235 isotope of uranium (German physicist Siegfried Fluegge had first calculated the energy balance of U238 and U235 fission and published it in June 1939 in *Die Naturwissenschaften*) and discovered that using merely a pound of it made building a bomb possible. In late 1943, Peierls led the British group that came to Los Alamos.

6. **EDWARD TELLER**

Edward Teller (b. 1908) lobbied President Franklin D. Roosevelt to build a nuclear weapon, and he was involved in the Manhattan Project from the beginning. One of his tasks was to calculate the possibility of a hydrogen bomb. Although initial calculations indicated that it was impossible, he soon realized that he was wrong and campaigned for its implementation even after the atomic bomb was dropped on Hiroshima. When the arms race began with the Soviets in 1949, Teller lobbied for a second nuclear weapons laboratory, established at Livermore, California. In 1956, he assured the navy that Livermore could build a warhead small enough to be fired from a submarine, and four years later, the *Polaris* submarine was armed with warheads. He later advocated a nuclear defense system powered by lasers.

7. HANS BETHE

Despite his wife's reservations, physicist Hans Bethe (b. 1906) accepted J. Robert Oppenheimer's invitation to join the Manhattan Project as director of the theoretical physics division. The project required him to explain how the atomic bomb would work and with what effect. Drawing on his knowledge of nuclear physics, shock waves, and electromagnetic theory, Bethe worked with Richard Feynman to devise a formula to calculate the efficiency of a nuclear weapon.

Born in Strasbourg, Bethe showed precocious talent in mathematics. Between world wars, Germany experienced extreme inflation. His father was paid twice a week to keep up with the devalued German mark. Hans was responsible for collecting his father's salary and spending it on food before the new value of the mark was calculated in the afternoon. After the war, the naturalized U.S. citizen advocated disarmament.

8. RICHARD FEYNMAN

Theoretical physicist Richard Feynman (1918–1988) worked on the Manhattan Project, developing a theory of predetonation that measured the likelihood that uranium might explode too soon. Later he found a new way to formulate quantum mechanics and quantum electrodynamics through path integrals. Feynman used a personal approach to understanding the physical laws and recognized the importance of visualization to understand physical phenomena. Feynman's diagrams help others visualize quantum electrodynamics. In 1965, he won the Nobel Prize for physics (with Shinichiro Tomonaga and Julian S. Schwinger) for his work on quantum electrodynamics.

9. **KLAUS FUCHS**

German physicist Klaus Fuchs (1911–1988) served as a member of the British team that worked with U.S. scientists to build the atomic bomb. In 1931, Nazi harassment led his mother to commit suicide. At 21, Fuchs was wanted by authorities for his political activities and fled to the West, arriving in England in 1933. Although he proved his worth to the British team by solving the equation concerning how much uranium was needed for the bomb, he soon began providing the Soviets with classified information. In December 1943, he became a member of the British mission to the Manhattan Project. After the war, Fuchs headed the theoretical physics division of the British bomb effort. When the FBI cracked the Soviet cipher code, it identified Fuchs as a spy. He was arrested by Scotland Yard in 1950 and served a nine-year prison term. After his release, he was named deputy director of the nuclear institute at Rossendorf and received many honors from East Germany.

10. **NORMAN R. DAVIDSON**

Organic chemist Norman R. Davidson (b. 1916) developed techniques for discovering what happens during rapid chemical reactions. His work was interrupted in 1942 by World War II. He joined the Manhattan Project and was assigned to research the production of transuranium elements at Columbia University and, later, at the University of Chicago.

GEOLOGY

Ancient Views on How Life Began

The concept of life emerging from a single basic substance is one of the recurrent themes in the myths of antiquity. Even though malevolent gods capriciously toyed with mankind, their priestly scribes and philosophers recorded "divine" activities. The distinctions between mythical and scientific approaches lie in the interpretation of these observations. Have we yet made the leap into totally rational thinking? And is such a leap even possible?

1. THALES OF MILETUS

Hailed as the "first man of science," Thales of Miletus (c. 624–547 B.C.) declared that the gods played no personal role in Nature, but he still identified the elements with the gods ("all things are full of gods"). He reasoned that all variations in Nature could be accounted for in terms of a single substance—water. Thales's pupil, Anaximander of Miletus (610–c. 547 B.C.), believed that the first principle was a boundless Apeiron that contained all the primary elements of matter: hot and cold, wet and dry. Through their intermingling, the world came into existence. The first living beings, he said, were generated in water, each enveloped with a "prickly bark." Adapting to land, they lost their protective covering and adapted to their new way of living. Man spontaneously appeared within a shark, nurtured within until he could survive on his own. Anaximander's pupil Anaximenes (c. 545 B.C.) urged that air was the primal substance, and all matter was a quantitive transformation of air, creatures being "homogenous air and wind."

2. XENOPHANES

Xenophanes (c. 560–c. 478 B.C.) adduced paleontological evidence to support the notion that earth and water are the stuff of life. The existence of fossils and seashells found inland could mean only one thing: The earth had once been covered by the sea, and life sprang from the mud and rocks when they emerged from the water.

3. HERACLITUS

A misanthropic hermit, Heraclitus (c. 500 B.C.) wrote *On Nature*. The first principle, he proclaimed, was neither water nor air but fire, for it embodied change. To the ancients, fire wasn't a chemical process but an element. Earth and water were fundamentally fire that had been extinguished.

4. EMPEDOCLES

Empedocles of Acragus (c. 490–430 B.C.) postulated four initial substances: fire, air, earth, and water. These four elements were set into motion by two basic forces—Love and Strife—the raw materials of the elements. He proposed four stages of evolution: (1) creation with unattached limbs; (2) the random union of limbs; (3) creatures being born with faces and breasts on both sides; and (4) parts separated from the whole. Change was still occurring. For example, even though the first animals were born with straight spines, some had curved spines because their ancestors had adopted the habit of turning their necks backward, and the acquired characteristic was passed on to offspring. Therefore, Empedocles has been called the father of the idea of evolution.

5. **THE ATOMISTS**

The Atomists, headed by Democritus (c. 460–c. 370 B.C.) and Leucretius (fl. 430 B.C.), taught that an infinite number of elements were always present and combined by chance to produce all other known substances. This was the high point of pre-Socratic philosophy.

6. **ARCHELAUS OF ATHENS**

Anaxagoras's (c. 500–c. 428 B.C.) pupil Archelaus of Athens (fl. 399 B.C.) was captivated by the similarities of Mother Earth to human motherhood. When life first emerged from the earth, a milklike fluid was produced to sustain the creatures. Because the slime didn't fulfill their nutritional needs, many died quickly, but survivors adapted by reproducing themselves.

7. **ARISTOTLE**

Aristotle (384–322 B.C.) believed that the soul determined form and function of an organism. Life didn't develop from lifeless matter but was present within it from the beginning of time—moisture and high temperature spawned it.

8. **THEOPHRASTUS**

After Aristotle left Athens, his pupil Theophrastus (c. 372–c. 287 B.C.) devoted himself to describing and classifying plants. He questioned the doctrine of abiogenesis, or spontaneous generation, but didn't dismiss it. He recognized that soil and climate affected plant growth.

9. LUCRETIUS CARUS

In *De Rerum Natura,* Roman poet T. Lucretius Carus (99–55 B.C.) proposed that the world originally consisted of nothing more than a vast expanse of randomly moving atoms. Some atoms joined for a short while, whereas others fused permanently. Then life appeared. Many species spontaneously emerged from the earth. While species themselves could not develop into other species while alive, this might occur following death as a result of the atomic patterns coming apart and reforming into other bodies.

10. DIODORUS SICULUS

In the first century B.C., Greek historian Diodorus Siculus (c. 90–21 B.C.) observed that the soil of Thebes generated huge mice, "some of them fully formed as far as the breast and front feet, and are able to move, while the rest of the body is unformed, the clod of earth still retaining its natural character." Diodorus concluded that when the world was taking shape, mankind came into being because of the "well-tempered nature of its soil."

Modern Views on Origins

The answers to two of the biggest questions on life's origins—How did Earth begin? and How did life begin?—have been hotly pursued and debated for centuries. Many say that life here began out there. One of the latest insights originates with two German scientists, Gunter Wachtershauser and Claudia Huber, who believe that life may have originated on the ocean floor from chemical reactions caused by molten lava.

1. GERMS

Benoit de Maillet (1656–1738), whose *Telliamed* (1748) provided the first theories of Earth's creation that disputed current theories, believed that generation occurs when a germ of the appropriate species finds its way into a female womb. Before there was life on Earth, germs matured in the vast ancient ocean. Original members of each species were formed through the development of germs that found their way into Earth. Germs also adopted to different conditions as they grew. When life first appeared on Earth, the planet was covered by water, and thus the first members of each species were aquatic. Stories of mermaids were not far-fetched, he believed.

2. **CLOUDS**

Fred Hoyle (1915–2001) and colleague Chandra Wickramasinghe (b. 1939) theorized that life evolved in interstellar clouds over a long period of time before the solar system ever formed. Doubling as a sci-fi author, Hoyle, in his novel *The Black Cloud,* speculated that life can exist anywhere when a flow of energy can be tapped. In Hoyle's book, this happened inside a cloud of interstellar material, making the cloud itself sentient.

3. **WARM, WATERY ENVIRONMENT**

In the 1920s, Alexander Oparin, a Russian biochemist, proposed the idea that life originated in the warm, watery environment of Earth's early surface, under an atmosphere mostly composed of methane. He believed that the early seas were rich in organic molecules, which formed more complex molecules until proteins and, thus, life were formed.

4. **PANSPERMIA**

Francis Crick (b. 1916), who shared a Nobel Prize with James Watson for the discovery of DNA, proposed that our galaxy was deliberately seeded with life by a civilization that arose billions of years ago when the galaxy was young.

5. **JUMP START**

At the University of Chicago in 1953, graduate student Stanley Miller mimicked Earth's primitive atmosphere in a laboratory. He enclosed methane, ammonia, and other gases in a glass flask partially filled with water. When he introduced electric sparks to imitate lightning, the clear water turned pink, then brown, as it became enriched with amino acids. Miller transformed origins-of-life research from speculation to experimental science. To explain how these building blocks linked into complex molecular structures, he speculated that organic molecules, floating in seawater, splashed into tidal pools, where they became concentrated through evaporation, like soup thickening in a pot.

6. SINGULARITY

Theorist Roger Penrose proved in the mid-1960s that matter that falls into a black hole must be crushed into a mathematical point at its center, known as a singularity. Turning this theory around, Stephen Hawking (b. 1942) showed that in the expanding universe, everything must have arisen from a singularity at the beginning of time.

7. FREE-LUNCH UNIVERSE

Could the universe have appeared out of nothing? In the words of Alan Guth, the father of the inflation model, this was the "ultimate free lunch." The idea developed from an unsigned commentary in a 1971 issue of *Nature* that suggested that the universe could be described as the inside of a black hole. Then, in a 1973 issue of *Nature,* Edward Tryon of the City University of New York suggested that a black-hole universe may have appeared out of nothing as a vacuum fluctuation, allowed by quantum theory.

8. ROCK AND ROLL

In 1998, a postdoctoral researcher, Jay A. Branies, and earth scientist Robert M. Hazen showed that the amino acid leucine breaks down within minutes in pressurized water at 200°C. When an iron sulfide mineral, commonly found around hydrothermal vents, was added, the amino acid stayed intact for days, time enough for it to react with critical molecules to form more complex structures.

9. MINERAL CATALYST

In 1998, Gunter Wachtershauser, a German patent lawyer with an abiding interest in life's origins, suggested that minerals, mostly iron and nickel sulfides that abound in deep-sea hydrothermal vents, could have been the catalyst for the formation of biological molecules. He theorized that primitive life existed as coatings that adhered to positively charged surfaces of pyrite (a compound of iron and sulfur). They obtained energy from the chemical reactions that produce pyrite. Indeed, some metabolic enzymes do contain a cluster of metal and sulfur atoms.

10. EKPYROTIC UNIVERSE

From the Greek word meaning conflagration, the *ekpyrotic* universe is thought to have been created from a huge fireball that cooled down after an unusual collision. Cosmologists ask, What caused the fire? The new theory has arisen from the superstring and M-theories that posit 11 dimensions. The three familiar ones of space, plus time, compose our universe, which stretches approximately 12 to 16 billion light-years across. Surrounding us, theorists surmise, is an oceanic fifth dimension, which might play a godlike role in starting everything in motion in our universe. Dimensions 6 through 11 provide the underpinnings of other dimensions.

Paul Steinhardt and Justin Khoury of Princeton University, Neil Turok of Cambridge University, and Burt Ovrut of the University of Pennsylvania theorize that there was a collision between "branes," or membranes in the dimensions. A huge, rogue wave, they say, rolled through the fifth dimension like a white wave runs through the ocean and destroys ocean liners. The wave moved into our four-dimensional space and started the big bang. The energy created by the collision set in motion quarks, leptons, and radiation, which, in turn, gave birth to heat, light, and matter.

How and When Earth Began

Current research has led to the theory that Earth is about 4.5 billion years old. Our view of Earth has shifted from the center of the universe to a mere planet formed by physical, not miraculous, forces.

1. RENÉ DESCARTES

To avoid criticism from the church, René Descartes (1596–1650), in his *Principles of Philosophy* (1644), suggested that Earth could have been formed from a star that cooled down to a ball of ash and then was trapped in the Sun's vortex. By suggesting a mechanical model, Descartes founded a significant trend affecting the early history of geology.

2. EDMUND HALLEY

In 1715, astronomer Edmund Halley (1656–1742), the discoverer of the eponymous comet, assumed that the oceans originally didn't contain any salt but were pure water that became salty over time as the result of saline particles transported by rivers. He thought that if the level of salt in the ocean was measured once, then again in another century, Earth's age could be ascertained. He lamented that no one had previously thought of this. But Halley's proposal wasn't heeded for another 180 years, until it was examined by John Joly, a professor of geology at Trinity College in Dublin in 1897. Joly arrived at an age of 89 million years since the formation of the first ocean.

3. WILLIAM WHISTON

Sir Isaac Newton tried to prevent his theories from being adapted to the Cartesian mechanical origins, but his objections were largely ignored. Newton's successor at Cambridge, William Whiston (1667–1752), suggested in *New Theory of the Earth* (1696) that Earth could have been created from a comet that condensed to form a solid body. The deluge was caused when another comet swept by Earth, depositing large quantities of water onto its surface.

4. COMTE DE BUFFON

French naturalist Comte de Buffon (1707–1788) suggested that Earth had been formed by the collision of the Sun with another massive body, which he called a comet. This theory was daring at the time—that a comet, not God, formed the planet. He was the first important scientist to suggest that Earth might be older than the 6,000 years allotted by Archbishop Ussher. Buffon also calculated that Earth was 75,000 years old, based on how long it would take Earth to cool from the temperature of the Sun.

5. LORD KELVIN

William Thompson, Lord Kelvin (1824–1907), professor of natural philosophy at Glasgow University, argued that mines and boreholes showed that the temperature within Earth increased with depth. From this he deduced that Earth was still cooling down from a time when it had originally been a molten globe. He believed that Earth's age could be calculated by figuring the rate at which rocks melted and cooled. Even though these values were unknown at the time, Lord Kelvin estimated that the consolidation could not have taken place less than 20 million years ago.

6. **SAMUEL HAUGHTON**

Irish geologist Samuel Haughton (1821–1897) proposed that the maximum thickness of strata is proportional to the time of their formation. Thus, the thicker the strata, the longer they took to form. Calculating sediments deposited on the ocean floor at the rate of 1 foot in 8,616 years, Haughton concluded that Earth must be 200 million years old. This was a gutsy proposal in the times when Lord Kelvin's ideas ruled, and it's closer to modern estimates.

7. **THOMAS CHAMBERLAIN**

In 1905, American geologist Thomas Chamberlain theorized that Earth had been created by the accumulation of cold, solid particles called planetesimals, opposing Lord Kelvin's assumptions that Earth had begun as a molten globe. Chamberlain said that a rapidly moving *star* passed close to the sun and the *gravity* of the star pulled arms of gas from the sun. The gas cooled and formed planetesimals, which collected in the centers of small eddies and, in turn, formed planets. Heat was dissipated in later stages, leaving Earth a cold body.

8. **ARTHUR HOLMES**

In 1946, Arthur Holmes (1890–1965), one of the foremost geologists of the 20th century, concluded that the probable age of Earth was approximately 3,350 million years. Thanks to the new Marchant Calculating Machine, he concluded in 1,419 solutions the time since the isotopic constitution of Earth's primeval lead began to be modified by the addition of lead isotopes generated from uranium and thorium. From radiometric data he concluded the age of uranium to be 4,460 million years. However, the lead sample Holmes used didn't contain the primeval lead value. Yet the mathematical model he developed is essentially the same one used today for calculating Earth's age. Holmes said at the end of his life, "The Earth has grown older much more rapidly than I have—from about six thousand years when I was ten to four or five billion years by the time I reached sixty."

9. CLAIR PATTERSON

According to Clair Patterson (b. 1922) of the California Institute of Technology in Pasadena, Earth was formed 4.56 million years ago. He reasoned that deep-sea sediments accumulated on the ocean floor, representing a wide volume of rock eroded from the continents. The sediments were thus likely to contain lead values representing the age of Earth's crust. He also analyzed the lead values of meteorites and decided that they had formed at the same time as Earth.

10. FIESEL HOUTERMANS

In 1953, Fiesel Houtermans, cofounder of the Holmes-Houtermans model for dating the age of Earth, estimated that Earth was 500 million years old, plus or minus 300 million years. He based his calculations on the age of meteorites that represented the age of primeval lead on Earth and the data from tertiary galenas.

What on Earth?

Scientific inquiry into geological change had been anything but scientific until the nineteenth century. Geology has been a science rooted deeply in controversy, divided on issues of questionable "facts" and religious dogma.

1. JAMES HUTTON

Trained as a physician, James Hutton (1726–1797) postulated a limitless geological time and agreed that the present is key to the past. He recognized the central role of Earth's internal heat in creating and transforming rocks and minerals. The life cycle of continents followed a ceaseless rhythm of erosion and deposition; continents were slowly worn down by running water and carried to the sea as sediment, later to be elevated as new continents. He concluded that a system of such balance and purpose had to emerge from an intelligent design. Volcanoes existed, he said, because their uplifting power was needed to raise the strata. However, Hutton's most ardent opponent, Irish mineralogist and chemist Richard Kirway, attacked his deism and said that it was a danger to revealed religion, which still revered Bishop Ussher's 4004 B.C. as the date of creation.

2. ABRAHAM GOTTLOB WERNER

Mineralogist Abraham Gottlob Werner (1750–1817) argued that Earth's strata had successively precipitated out of solution from an original ocean. Basalt, he went on, was not a volcanic rock but chemically precipitated sediment. He and other "neptunists" advocated a theory of Earth history in which a primeval ocean formed a geological record as we see it. This theory had the theological advantage that it seemed to accord with the biblical flood, while the geologic logic stemmed from the theory that sedimentary beds were created and cemented at the bottom of the sea. Werner's detractors claimed that he traveled too little and was beset by such a passion for order that his theories were too determinate. Werner did devote much time to preparing seating charts for dinners and also arranging and rearranging his library. He believed that well-trained senses could function without instruments or experimental techniques to explain what we think are mysteries.

3. SIR CHARLES LYELL

To Sir Charles Lyell (1797–1875) we owe *Principles of Geology,* which Darwin credited as an inspiration for his vast time scale. He believed that the present is the key to the past—the same agencies that change Earth now changed it in the past, and with similar intensity of action. Lyell's theory, called uniformitarianism, claims that the chief forces at work are erosion and sedimentation, the action of earthquakes and volcanoes, and the rise and fall of land and sea levels.

4. ÉLIE DE BEAUMONT

French geologist Élie de Beaumont (1798–1874) began as a gifted mathematician and studied geology as it related to mines. His mapping of the mountainous eastern portion of France and surveys of mineral resources brought him a good general view of European geology. He believed he could apply mathematical techniques to a complex set of phenomena such as the uplift of mountains. Mountains were arranged, Beaumont argued, in distinct areas as a result of a simultaneous uplift along "Earth's great circles." He granted that erosion, sedimentation, volcanism, and plutonic injection played a part in shaping Earth's surface, but he also allowed catastrophes to have their place. Mountains, Beaumont reasoned, were not the result of earlier paroxysms. He suggested that Earth underwent periodic readjustments in which the crust diminished its circumference by thrusting up, or collapsing, along great circles, in order to remain in contact with a shrinking interior.

5. **EDUARD SUESS**

Austrian geologist Eduard Suess (1831–1914) began geological work at age 19, but, because he joined the revolutionary ferment of 1848, he was imprisoned. His jailing and political record knocked him out of admission to school. Suess began his geological career as a clerk in the paleontological section of the Imperial Geological Museum. He had studied the Alps, which he thought were the "type range" for folded mountains and believed that their structure could solve the problems of dynamic geology. He said that all the major ranges of Europe had experienced a general northern displacement together and moved as a system, advancing forward imperceptibly, as slowly as ocean waves. He also said that the principal problem faced by 19th-century geology was a limitation of perspective.

6. **MARSHALL B. GARDNER**

Marshall B. Gardner (1854–1937) worked as a foreman in a corset factory in Aurora, Illinois, but he also lectured on his favorite theory—that Earth was hollow. He said that beneath an 800-mile-thick crust was a small sun, 600 miles in diameter, that provided 24 hours of daylight. From openings at the 1,400-mile-wide poles came the aurora borealis. Gardner claimed that his theory was based on solid fact.

7. JOHN JOLY

In 1909, Irish physicist John Joly (1857–1933) published *Radioactivity and Geology,* introducing the idea that radioactivity provided many answers to geology's mysteries. Rocks in Earth's crust had been found to be radioactive. This allowed him to strike down calculations of Earth's age based on contraction. The amount of heat produced by radioactive decay, said Joly, showed the rate at which Earth might cool. He even suggested that Earth may be heating up. He believed that continents were permanent features of the crust and had been growing since the Paleozoic era by accretion of infilling geosynclines that were uplifted into mountains and eroded seaward.

8. ALFRED WEGENER

Trained as an astronomer, German geophysicist Alfred Wegener (1880–1930) practiced meteorology while also fulfilling his love of exploring the world, participating in several Arctic expeditions to the Greenland ice cap. Studying many maps, he was struck by the similarity of opposing coasts abutting the Atlantic Ocean. In *Origin of Continents and Oceans,* published in 1912, he argued that one massive continent, Panegea, had been rent into parts that drifted. The largest piece, Gondwanaland, consisted of South America, Africa, Antarctica, India, and Australia. This landmass traversed the South Pole during the Paleozoic era. As Gondwanaland moved northward, it encountered Larasia. Today, Wegener's hypothesis of continental drift prefigures the theory of plate tectonics.

9. HARRY HAMMOND HESS

In 1960, American geologist Harry Hammond Hess (1906–1969) proposed the theory of seafloor spreading. The ocean floor, he claimed, is constantly being created at underwater ridges in the middle of the oceans, spreading outward, and being consumed in trenches underneath the continents. In the mid-1960s, new data on magnetic anomalies in the Pacific Ocean revealed that seafloor spreading does occur. This mechanism was considered proof of Wegener's continental drift theory.

10. JOHN ROGERS

In 1996, John Rogers, a University of North Carolina geologist, published a paper entitled "A History of Continents in the Past Three Billion Years." While working in India, he studied a craton, the ancient core of a continent, this one being approximately 3 billion years old. When he studied other cratons, he theorized that they belong to a 3-billion-year-old continent, the first one on Earth, which he called Ur. For 500 million years, Ur wandered alone, growing slowly as volcanoes spewed magma. By 2.5 billion years ago, a new continent appeared that Rogers called Arctica (Canada, Greenland, and most of Siberia). Then, 2 billion years ago, Baltica (western Europe) and Atlantica (eastern South America and western Africa) arose. Approximately 1.5 billion years ago, Baltica and Atlantica collided, forming Nena. One billion years ago, Ur and Atlantica plowed into Nena, producing the supercontinent Rodinia, which survived 300 million years, and then fell apart into Ur. Rifts formed within the original cratons, scattering the present continents across the oceans.

How Old Is Earth?

We have tried for centuries to pinpoint when Earth was formed. Sir James Ussher (1581–1656), Archbishop of Armagh, decided in 1650 that Earth had been created 4,004 years before the birth of Christ, on the evening of October 22, a Saturday. This was accepted in Christian teaching for centuries. Other religions had different ideas. Zoroaster, the sixth-century prophet, believed the world had been in existence for more than 12,000 years. Against the background of dogma, it was difficult to cram millions of years of Earth's history into a few thousand years. Let's discover how scientists attempted the feat.

1. COMTE DE BUFFON

Biblical chronology went virtually unchallenged until French naturalist Comte de Buffon (1707–1788), in 1749, tried to determine Earth's age by scientific reasoning. He said that Earth was created approximately 75,000 years ago, based on his understanding of Earth's internal heat and rate of cooling.

2. JAMES HUTTON

In 1785, James Hutton (1726–1797), frequently called the father of modern geology, presented his "Theory of the Earth," in which he emphasized the immensity of geological time and the uniformity of geological processes. He claimed, "We find no vestige of a beginning and no prospect of an end." Hutton was accused of "deposing the Almighty Creator of the Universe from His Office."

3. CHARLES WALCOTT

In 1893, Charles Walcott (1850–1927), who later became secretary of the Smithsonian Institute, published a thorough study of accumulated sediments. By adding the thicknesses of strata of various ages, he guessed that geologic time encompasses 55 million years.

4. LORD KELVIN

In the late nineteenth century, British physicist William Thomson (later Lord Kelvin, 1824–1907) countered geologists' opinions by arguing that Earth was young. He thought there could have been only one source of energy—gravity. Thus, the planet began as a cauldron but cooled at a constant rate. To arrive at Earth's age, Kelvin calculated the time necessary for Earth to cool to its present condition. He settled on an age of 100 million years, which he revised over the next three decades to 2.4 million years.

5. JOHN JOLY

In 1899, Irish mineralogist John Joly (1857–1933) reasoned that rivers, in addition to carrying sediments, must deliver salt from the continents to the sea. Assuming that no salt was ever lost from the oceans, he calculated that 90 million years were required to account for their salinity.

6. ERNEST RUTHERFORD

In 1903, New Zealand–born physicist Ernest Rutherford (1871–1937) and British chemist Frederick Soddy (1877–1956), then both faculty members at McGill University, demonstrated that radioactive atoms decay spontaneously to a more stable form and release heat. This cast doubt on Lord Kelvin's chronology of Earth's age at 2.4 million years, based on a cooling planet. Those who had argued from geological evidence that Earth must be far older felt vindicated. However, it turned out that radioactive heat generated inside Earth is not enough to change the cooling time significantly. Kelvin's fundamental error lay in his assumption that Earth is cooling by conduction, when radiation is actually at work. Also, Rutherford found that a uranium-rich mineral from Connecticut must have been in existence for at least 500 million years to account for its concentration of helium. Cambridge physicist R. J. Strutt, using the decay of radioactive thorium, showed a minimum age of 2.4 billion years. However, scientists learned that helium cannot be relied on because it leaks out of crystals.

7. **ARTHUR HOLMES**

British physicist Arthur Holmes (1890–1965) was convinced that radioactivity could provide geology with a precise clock. He believed that lead was produced by radioactive decay, so that when Earth formed, it must have contained little or no lead. In 1927, by estimating the amount of lead and uranium in Earth's crust, he arrived at an age between 1.6 and 3 billion years.

8. **CLAIR PATTERSON**

A graduate student at the University of Chicago, Clair Patterson (b. 1922) measured the minute quantities of lead that occur in rocks and minerals, rather than just lead ores. He used the mixture of lead isotopes in sulfides of a meteorite, which contained no radioactive source that could produce new lead, as his beginning point. He then used lead isotopes in sediments deposited on the ocean floor to calculate their age. His samples pointed to an age of 4.5 billion years. Patterson's paper, published in 1956, refined the age to 4.55 billion years.

9. STEPHEN MOORBATH

In 1971, Stephen Moorbath, a geochronologist at Oxford University, was studying the ancient rocks of Greenland when a mining company prospecting for iron in the Isha Hills alerted him about an unusual finding. When Moorbath visited the region, he discovered fine-grained sedimentary rocks that had tubular and pillow shapes, typical of lavas that erupted under water. Moorbath dated them using the rubidium-strontium method of radioactive dating. This yielded a staggering age of 3.75 billion years, making it the oldest part of Earth's surface ever discovered.

10. JOHN BAUMGARDNER

John Baumgardner, a Los Alamos scientist and the preeminent expert in the design of computer models for geophysical convection, created Terra, a program that can produce three-dimensional maps of the Earth's mantle convecting through time. However, he believes that the program also proves the biblical flood and that Earth was created by God less than 10,000 years ago. The cosmos itself, insists Baumgardener, is only a few thousand years old. Baumgardner uses the theory of "runaway subduction," which posits that gravity pulls the heavy plates of the ocean floor under the continents into the mantle, which is silicate rock, which, when heated, sinks faster.

Ice Age Theories

The theory of ice ages was first proposed in the nineteenth century by, among others, Swiss civil engineer Ignace Venetz in 1821 and Swiss naturalist Louis Agassiz in 1847. Before then, most geologists had believed that rocks and sediment had been molded by the biblical flood. Botanist Karl Schimper (1803–1867) first used the term *ice age* in 1837.

The cause of ice ages is a remarkably contentious topic, but there are some points that most researchers accept. Sixty million years ago, Earth started to cool in earnest. Twenty-five million years later, Antarctica was buried under a thick sheet of ice. Then, 16 million years after that, glaciers overran Greenland. Finally, ice sheets invaded North America and Europe, retreated, advanced, retreated, and so on.

1. **AXIAL PRECESSION**

In the 1860s, Scotsman James Croll (1821–1890) showed that the change in Earth's orbit from high to low eccentricity, and back again, takes about 100,000 years. Once in that period, Earth is unusually far from the Sun. What if, Croll wondered, the axial precession (the "wobble" of Earth's axis that forms an imaginary circle in space every 22,000 years) and the 100,000-year orbital cycle were combined? He believed that these conditions created ice ages, or "global epochs," that lasted 10,000 years.

2. MILANKOVITCH MODEL

A Yugoslavian climatologist, Milutin Milankovitch (1879–1958) developed the theory that the cycle of ice ages on Earth is caused by changes in Earth's orbit that alter the balance of heat among the seasons. Although the total amount of heat derived from the Sun in one year is always the same, sometimes there is a large difference between summer and winter. The effect produced rhythms roughly 26,000, 40,000, and 90,000 to 100,000 years long.

3. HEAVY OXYGEN

A geophysicist at the University of North Carolina at Chapel Hill, José Rial used frequency modulation, which imposed the shape of a long-wave signal frequency onto a short-wave "carrier" frequency, to analyze the amount of "heavy" oxygen of deep-sea sediments over the past million years. Water containing heavy oxygen has two more neutrons in its nucleus than the common form of the elements and does not evaporate as quickly. During periods when an ice sheet grows on land, the oceans are depleted of light oxygen. When ice sheets shrink, melted water returns to the oceans, and the proportion of light oxygen increases again. Rial figured that this would be reflected in oxygen isotope ratios in fossil shells. He calculated that ice ages have occurred every 100,000 years, but with a variation between 80,000 and 120,000 years. He feels that this substantiates the Earth-orbit theory. Oceanographers James Hays, John Imbrie, and Nicolas Shackleton also conducted similar tests in 1976, finding periods of 41,000 and 23,000 years.

4. CLIMACTIC BUZZ

Geographer–plant ecologist Katherine Willis of Oxford University and her University of Cambridge colleagues suggested in a 1999 article in *Science* that the ice age that occurred 2.75 million years ago was started by the intensification of one orbital cycle, which produced a climactic "buzz." Studying the sediments of a now-vanished lake in central Hungary, Willis sampled the core at 2,500-year intervals, extracting pollen whose species composition varies as the climate changes. The pollen shows short warming and cooling cycles lasting 5,000 to 15,000 years, which are shorter than any astronomical cycle.

5. EXTRATERRESTIAL DUST

Extraterrestial dust, not periodic flutters in Earth's orbit, caused the last 10 ice ages, according to physicist Richard A. Muller of the University of California at Berkeley and Gordon J. MacDonald of the International Institute for Applied Systems Analysis in Laxemburg, Austria. In 1993, they suggested that when Earth's orbit reaches a certain plane, the planet plows through a thick cloud of interplanetary dust. The small particles drift through the upper atmosphere and block out sunshine, stimulating cloud growth or weakening the ozone layer. Thus, Earth cools enough to produce an ice age.

6. THE HIMALAYAS

The cooling that started 60 million years ago was caused by a drop in carbon dioxide levels in the atmosphere. Why did they fall? When in graduate school, marine geologist Maureen Raymo, now at the Massachusetts Institute of Technology, believed levels dropped because two of Earth's tectonic plates collided to form the Himalayas, the highest mountains on Earth. When carbon dioxide from the atmosphere combines with rain to make an acid that erodes rock, minerals such as calcium silicate in the rocks react with carbon dioxide, removing the gas from the atmosphere. Therefore, Raymo says, erosion on slopes as vast as the Himalayas could reduce carbon dioxide levels enough to give the ice age a push.

7. CLIMATE OSCILLATION

After analyzing a sample from Devil's Hole, Nevada, that was drilled from a wall of calcite, hydrologist Isaac Winograd of the U.S. Geological Survey in Reston, Virginia, says that he finds no support for the orbital theory. He claims that ice ages are not caused by external factors but by an internal oscillation in the climates. Close examination of the calcite core, dated by radioactive techniques, suggests that the climate was warming 140,000 years ago, and the ice was retreating rapidly.

8. POPSICLE PLANET

During the Neoproterozoic glacial period, Earth was covered entirely with ice and snow, said Joseph L. Kirschvinik of the California Institute of Technology. Almost all land areas were plastered together to form the supercontinent Rhodinia, which was centered in the tropics, leaving a single vast ocean. The globe, he said, would have been a "snowball Earth," with pack ice covering much of the ocean. If this happened, all life would have perished. This could have been set off by plate tectonics, plus a dimmer Sun, which was about 7% dimmer than it is today.

9. PLANT-ANIMAL RATIO

At least some cooling periods can be attributed to the evolutionary competition between land plants and herbivores, says paleontologist Paul Olsen of the Lamont-Doherty Earth Observatory in New York. The frost line has surged back and forth over the past 450 million years, dragging the climate along with it by changing the amount of carbon dioxide in the atmosphere and putting it in the ocean, diminishing the greenhouse effect and cooling Earth. Dinosaurs, the greatest plant eaters that ever lived, changed the ecosystem back to warmer days. However, the impact of a comet or asteroid caused mass extinction on both sides, and the cooling continued for millions of years because the angiosperms and grasses thrived.

10. **MUD SLIDE**

The effects that can shape Earth's climate are often subtle, or at least unexpected. In 1998, a team from the Southhampton Oceanography Center in Britain announced in *Nature* the discovery of a 450-billion-cubic-meter deposit of mud on the seafloor off the coast of Sardinia—the aftermath of a mud slide that would engulf all of France to a depth of almost a meter if it happened today. Carbon dating placed the slide at 20,000 years ago, at the height of the last ice age. This was a time when so much water had been turned to ice that sea levels were 120 meters lower than they are today. Formed from thousands of years of river deposits, the mud would have been rich in organic material that would produce methane gas when decomposed. Normally, this gas would have been locked into the mud by the huge pressure of seawater lying over the sediments. But as the ice age deepened and sea levels fell, the mud deposits would be exposed, and thus able to release their pent-up methane gas. The sudden release of methane might have helped push Earth out of the ice age. Researchers at Duke University in North Carolina argue that the sheer weight of ice pushing down on Earth's massive submarine mud slides would release the methane.

Mineral Monikers

As the mineral world expands, scientists must search for names for new additions. Some take the easy route and name discoveries after themselves. Others take a tougher approach, inventing creative names.

1. AMETHYST

The purple variety of quartz was named from the Greek word for "not drunk." The mineral was thought to counteract drunkenness.

2. ARMALCOLITE

This gray, metallic, moon mineral is attributed to three American astronauts on the *Apollo XI* lunar mission in 1969: Neal Alden Armstrong, Edwin Eugene Aldrin Jr., and Michael Collins.

3. ASBESTOS

Some people call it "powdery death," while some contractors curse its existence. But the general term of asbestiform minerals is Greek for *inextinguishable,* after the misconception that once lit, the material could not be extinguished.

4. BUNSENITE

German chemist Robert Wilhelm Bunsen (1811–1899) developed spectrum analysis in 1859; discovered cesium and rubidium; and improved, but did not invent, the gas burner in 1855, now called the Bunsen burner. Credit for the current design and manufacture of the burner goes to Peter Desaga, a technician at the University of Heidelburg, who designed it with Bunsen's specifications in mind.

Although Bunsen's work brought him wide acclaim, he nearly died of arsenic poisoning. The compound also cost him the sight in one eye when an arsenic explosion sent glass everywhere.

5. COLUMBITE

The dark mineral uses the old name for the United States—Columbia. Columbite has shown superconductive properties and is being studied with other metals for a possible breakthrough alloy in this industrial field. The official name became niobium in the 1950s after a century of debate, although some groups still do not recognize the official name and refer to it as columbite. Most geologists still refer to its namesake mineral as columbite instead of the proposed "niobite."

6. GARNET

This large group of minerals was named from the Latin *granatum* for "pomegranate," the seeds of which it was thought to resemble.

7. JOESMITHITE, JIMTHOMSONITE

Although not the discoverers, a vast number of mineralogists had minerals named for them. This list includes many elected presidents of the Mineralogical Society of America, such as James B. Thompson Jr., who served in 1968, and Joseph V. Smith, who served in 1973.

8. SKLODOWSKITE

Marie Curie's contributions to science deserved a mineral recognition, but curite was already named for her husband, Pierre. Instead, her maiden name, Sklodowska, was used.

9. SMITHSONITE

James Smithson (1765–1829), the British chemist and mineralogist who studied the chemistry of minerals, never visited the United States, but left his estate to a nephew, saying that if the nephew died childless, the estate would go to America to found the Smithsonian Institute (which occurred in 1846).

Causing many debates on both sides of the Atlantic, President Andrew Jackson asked Congress to pass legislation allowing him to accept the gift, unsure whether the Constitution gave him this authority.

Senator John C. Calhoun, among others, opposed acceptance and maintained that Congress had no authority to accept the gift. Congress finally authorized acceptance in 1836, and a museum was agreed on 10 years later.

10. TRANQUILLITYITE

Along with armalcolite and pyroxferroite, found only on the moon, this mineral comes from the Sea of Tranquillity and was taken during the *Apollo XI* mission.

BIOLOGY

Aliens in Our Backyard

Invasive plants carry similar qualities from species to species, each quality helping to overtake an ecosystem's natural plant, vertebrate, and invertebrate inhabitants. Early maturation, long life, seed overabundance, and high photosynthetic rates all help trespassers get the advantage over residents.

If you think the number one interloper here should be the dandelion, we would agree. But taking out such an obvious choice, here are 10 runners-up for most invasive nonnative plant.

1. WATER CHESTNUT: A THORN IN THE SIDE OF NORTH AMERICA

Trapa natans

Although a native of Asia, Africa, and Eastern Europe, this intruder is not the crunchy edible plant used in Asian cooking—a shame, because the way that this virile plant spreads, it could feed the world.

The water chestnut was first recorded in North America near Concord, Massachusetts, in 1859, and later introduced to Collins Lake, New York, in 1884 as an ornamental water plant. Since then, the unwelcome guest continues to spread throughout New England and the Mid-Atlantic states. The annual, fast-growing, floating aquatic grows up to 16 feet long. Mats of featherlike submerged leaves and triangular jagged-surface leaves choke waterways for boating, fishing, and swimming; monopolize light; and increase sedimentation, but the naughty bits are the thorny black nutlets that hold painful terminal barbs capable of piercing shoe leather. The barbs easily disperse by water in late July.

2. **KUDZU: FRIEND OR FOE**
Pueraria montana, lobata, triloba, thunburgiana

A lot of man-hours and sweat helped kudzu become as prolific as it is today.

The Japanese government introduced kudzu, a pea-and-bean family member native to China, Taiwan, Japan, and India, to the United States in 1876 at the Centennial Exposition in Philadelphia, Pennsylvania. Unfortunately, its natural insect enemies were not brought with it.

By 1900, kudzu was available through mail order as inexpensive livestock fodder. In the 1930s, the Soil Conservation Service distributed approximately 85 million seedlings to Depression-desperate workers to control erosion. The government offered up to $8 per acre if farmers planted kudzu fields. The 1940s saw kudzu clubs and the crowning of kudzu queens at kudzu festivals.

By the early 1950s, kudzu had swamped everything it could grow on, but it wasn't declared a weed by the U.S. Department of Agriculture until the 1970s. The government estimates that kudzu now covers 7 million acres in the southeast, and it can be found as far north as Connecticut and west to Texas and Oklahoma. Mississippi, Alabama, and Georgia claim the largest infestations. Friends of the plant still fight negative publicity, touting baskets, paper, kudzu-blossom jelly, syrup, fried kudzu leaves, flour, and possibly a cure for alcoholism as some of its virtues.

3. JAPANESE HONEYSUCKLE: SWEET SMELL OF INVASION

Lonicera japonica

Possibly a black widow of the plant world, Japanese honeysuckle can topple trees and shrubs with its perfumed flowers and luscious green vines. As honeysuckle invades forest openings, it blankets trees, shrubs, and herbs, leaving sunny holes for itself and other nonnative species. *Lonicera japonica,* indigenous to east Asia and not to be confused with native strains, grows during part or all of the winter, giving it a further competitive edge.

Japanese honeysuckle was introduced on Long Island, New York, in 1806, then took 100 years to become established over the eastern United States. Originally introduced as a landscape plant, honeysuckle is still considered a desirable species by some landscapers, highway designers, and wildlife managers for its fragrant flowers, rapid growth, erosion control, bank stabilization, and winter forage. A mixture of honeysuckle parts and other plants reportedly has antibiotic and antiviral qualities, while leaves and flowers are used to heal chicken pox.

4. SALTCEDAR, OR TAMARISK: A PINCH OF SALT

Tamarix spp.

Saltcedar isn't only an aggressive, woody plant that has become established over as much as a million acres in the western United States. The gate-crasher is actually a soil pollutant too. Saltcedar concentrates salts in its leaves, which then are deposited on the soil when the leaves drop and disintegrate every season. Over time, surface soils become highly saline, impeding native plant growth. Native to Africa, Asia, and Europe, saltcedar is a relatively long-lived plant that tolerates a wide range of environmental conditions and produces massive quantities of small seeds.

In the early 1800s, eight species of saltcedar were introduced into the United States from Asia as ornamentals, as windbreakers, and to combat erosion. Three of the species invaded the southwestern United States. By 1961, at least 1,400 square miles of the western U.S. floodplains were infested, and it now occupies more than 1 million acres west of the Great Plains, north into Montana, and south into northwestern Mexico.

5. MILE-A-MINUTE: NEED WE SAY MORE?
Polygonum perfoliatum

A northeasterner's kudzu, the prickly annual vine from Asia has pale green triangular leaves and blueberry-like fruit. Typical habitats are roadsides, forests, stream banks, low meadows, orchards, and nurseries.

Mile-a-minute was first collected in the United States from ship ballast near Portland, Oregon, in the 1890s. The plant next appeared in rhododendron nurseries in York County, Pennsylvania, in 1946. Since that time, mile-a-minute has spread to New England and Mid-Atlantic states and has reached as far south as Mississippi.

Since its first appearance, the invader has been spread by birds and rodents and carried in rivers and streams. Under favorable growing conditions, mile-a-minute spreads rapidly in a niche similar to Japanese honeysuckle. Like many, it climbs like a mountain goat and can cover shrubs and trees.

6. LEAFY SPURGE: THE SCOURGE OF THE NORTH
Euphorbia esula

Leafy spurge, native to Eurasia, is one of the most serious weeds in the northern United States, causing millions of dollars in crop and ranching losses and control costs.

A deep-rooted, perennial rouge, leafy spurge hitched a ride into the United States in the 1830s with other seeds. It now infests about 2.7 million acres in southern Canada and the northern Great Plains of the United States. In North America, it forms dense groves that displace natural vegetation, causing loss of plant diversity, wildlife habitat, and land value. The government reports that cattle refuse to graze in areas with 10 to 20% leafy spurge cover, because the sap irritates cattle's digestive tract and causes lesions around the mouth and eyes. In the United States, direct livestock production losses together with indirect economic effects due to this species alone approached $110 million in 1990.

7. **PURPLE LOOSESTRIFE: MASKED MAGENTA MARAUDER**
Lythrum salicaria

A perennial herb that grows up to 8 feet tall, purple loosestrife was introduced into the United States from Europe in the early 1800s in ship ballast, as a medicinal herb and as an ornamental plant. Since 1880, purple loosestrife has increased rapidly, clogging waterways and drainage and displacing native vegetation. The plant now grows wild in at least 42 of the 50 states, with the greatest concentrations in New England, Great Lake, and Mid-Atlantic states. It grows best in freshwater marshes, open stream margins, and sandy floodplains, similar to areas that would have cattails and reed canary grass.

One would think people would exterminate these prolific weeds like cockroaches. However, nurseries across the country still sell purple loosestrife as an ornamental. It is promoted for use as a landscape plant and as a nectar plant in honey production. About 24 states have listed purple loosestrife as a noxious weed and prohibit its sale and distribution. Purple loosestrife is very difficult to control once established.

Purple loosestrife (center) was introduced into the United States in the early 1800s and is very difficult to control once established. Nevertheless, it is often promoted for use as a landscape plant, although 24 states currently prohibit its sale and distribution.

8. HYDRILLA: IT COMES IN PEACE AND INVADES IN PIECES

Hydrilla verticillata

Hydrilla, a submerged perennial, roots in freshwater lakes, streams, and riverbeds until it breaks free and forms floating mats, choking bodies of water. The fiend is native to Asia but has spread all over the world, possibly through a discarded aquatic fish hobby. In the United States, hydrilla occurs in all of the Gulf and Atlantic Coast states as far north as Connecticut, and on the West Coast in California to Washington.

In 1947, an aquatic plant dealer in St. Louis, Missouri, imported hydrilla from Ceylon. There is also note of it appearing in the Tampa, Florida, area in the early 1950s for use as an aquatic ornamental.

Free hydrilla was discovered in the United States in 1960 at two Florida locations, a canal near Miami and in Crystal River.

In "The Perfect Aquatic Weed," Kenneth Langeland states that there is evidence of at least two hydrilla introductions into the United States, because at least two different forms occur. Langeland also says that the growth habits of hydrilla enable it to compete so effectively. It elongates quickly, up to 1 inch per day, until it nears the water surface, where it branches out to steal the light from other submersed aquatic plants.

9. QUEEN ANNE'S LACE: NOT A BAD-LOOKING WEED
Daucus carota

What came first, the carrot or the lace? Queen Anne's Lace, of the carrot family Umbelliferae, is a native to the Mediterranean region but is found as a weed throughout North America. Similar to the cultivated carrot, it has feathery foliage and a woody root. Tiny white flowers form a flat cluster called an umbel, thus the family name. When they wither, the cluster becomes nest shaped. The plant was formerly used in folk medicine as a diuretic and a stimulant.

A commoner in fields and along roadways, Queen Anne's Lace is said to have edible roots. In its first year, the biennial plant grows only a small, tough root. In its second year, the white root is up to 6 inches long. The root is edible when scraped and boiled, with a smell identical to that of garden carrots. Some debate whether the plant came from domestic carrots gone to seed, or our garden carrots are derived from this plant.

10. ENGLISH IVY: A QUIET MENACE
Hedera helix

Quietly creeping up trees into the canopy with small rootlike hands, English ivy exudes a gluey substance to seal itself to its host. Dark green, heart-shaped, waxy leaves form a thick canopy to prevent sunlight from reaching other plants. Vines climb tree trunks and surround branches and twigs, blocking sunlight from the host tree to slowly kill the surrounding plants. The vine's added weight causes trees to break or blow over during storms. The plant flowers and forms berries, which birds eat and disperse to other locations.

A shade lover, English ivy is a ginseng family member native to Europe, western Asia, and northern Africa. It was introduced as an ornamental into North America from Eurasia during colonial times. The popular plant is recommended as a low-maintenance alternative to grass lawns, but the plants have outwitted the less avid gardener and grow abundantly in at least 26 states, especially Washington and Oregon, where it is one of the most abundant and widespread invasive plants.

Invasive Animals

Not as abundant as invasive plants, several groups of animals are still as detrimental to native species. Oftentimes while reading or watching the news, you can't help but think "whoops!"

1. **ARGENTINE ANTS**

Along with the southern United States's fire ants, the Argentine ants *(Linepithema humile)* have spread and become a nuisance throughout the world. The species establishes colonies with multiple queens and competes with a variety of other insects and some birds, as well as feeds on crops. One example: reaching South Africa in freight and baggage, the ants have spread and endangered the red stump plant, whose seeds would have been dispersed by local ants.

2. **AFRICANIZED BEES**

Called killer bees by the media, Africanized honey bees are the same species as European honeybees, which are used to produce honey and pollinate crops. They are the result of deliberate interbreeding between European bees and bees from Africa and were released in Brazil in the 1950s.

Africanized bees defend their colonies much more vigorously than do European bees. The colonies are easily disturbed, and when the bees do sting, many more bees may participate, so there is a danger of receiving more stings. Once disturbed they will continue the attack for a long distance. They also nest in places European bees do not, including small cavities near the ground such as water meter boxes or overturned flowerpots.

3. GRASS CARP

Also known as the white amur, *Ctenopharyngodon idella* is a vegetarian minnow-like fish native to the Amur River in Asia. Because of its feeding habits, the fish can be used as a biological tool to control nuisance aquatic plant growth, but it is mostly only dispersed in its genetically altered, sterile, triploid form. The fish grow rapidly, and they have been known to reach a weight of 100 pounds or more and live up to 15 years. Their maximum length ranges from 3 to 5 feet. Many states require a permit to transport the fish, and their reproduction is monitored closely.

4. ZEBRA MUSSELS

First discovered in the United States in 1988, Zebra mussels (Dreissena Polymorpha) clog water intake pipes in many cities along the Great Lakes and the Ohio and Mississippi Rivers and have endangered other mollusks, fish populations, and aquatic species.

The small, fingernail-sized freshwater mollusks colonize on surfaces, such as docks, boat hulls, commercial fishing nets, water intake pipes and valves, native mollusks, and other Zebra mussels, causing enormous economic and ecological effects. They feed voraciously and hoard the food that they cannot immediately consume by binding it with mucus, thereby endangering all the species that feed above them.

Zebra mussels mainly are transferred when they attach themselves to boats and trailers and can survive several days out of water. They also live in a free-swimming larval stage for one or two weeks and can be easily transferred to other places during this time.

5. ICHNEUMON WASPS

A 4-inch stinger on an inch-long wasp is the stuff of nightmares for wood-boring grubs in Appalachian forests. The "stinger" is actually an ovipositor that lays an egg inside the grub and consumes it from inside.

Ichneumon wasp larvae nibble on the grub's nutrient reserve, a store of energy used to fuel metamorphosis into the adult life. The parasitic maggot will devour these nutrients and use them to fuel its own growth into a wasp. Some Ichneumon eggs hatch when their host goes into the dormant pupal stage; other eggs hatch shortly after they are laid and get straight to work.

6. COWBIRDS

Blessing only North America with their presence, cowbirds *(Molothrus ater)* use other bird species as hosts. Females lay their eggs in other's nests and rely on these hosts to incubate and raise their chicks, called brood parasitism. The cowbird's clutch size is usually four to five eggs, with an incubation period of 10 to 13 days, usually faster than the host's eggs.

Brown-headed cowbirds have parasitized more than 220 host species in the United States, including the wood thrush, blue-winged teal, and red-headed woodpecker. While not all hosts make good foster parents—a number of species reject cowbird eggs—cowbird chicks have been successfully reared by more than 150 host species, with songbirds comprising the majority of the hosts.

7. STARLINGS

The European, English, or common starling *(Sturnus vulgaris)* is another invasive species that takes over nest boxes from native species. English starlings were introduced into the United States between 1850 and 1900. The first successful introduction consisted of approximately 60 European starlings released in Central Park, New York, in 1890.

Now ranging from coast to coast, the starling is a fierce competitor with native species. Starling nests are removed from bluebird habitats to allow more nesting sites for bluebirds, but other birds such as the red-headed woodpecker, flickers, and crested flycatchers have felt their presence.

Another negative aspect of this species is its habit of roosting in large numbers. Starlings constitute a nuisance when they roost because of the noise produced and the damage caused to public and private property by their droppings.

8. BROWN RAT

At the start of the eighteenth century, the European black rat *(Rattus rattus)* was the carrier of fleas that caused bubonic plague. The plague originated in the steppe lands of Eurasia, then migrated to Europe in the fourteenth century. After 1600, the brown rat *(Rattus norvegicus)* began to replace the black rat in Europe. The brown rat was not nearly as hospitable to fleas, and the plague began to subside, resulting in an increase in the human population in the eighteenth century.

9. FERAL HORSE

Feral horses *(Equus caballus)* are the most widely dispersed of equids, with populations found throughout the world. In North America, domestic horses, taken from Eurasian populations of wild horses, were introduced into most parts of the continent in the eighteenth

The *Rat Catcher* by Cornelis Visscher. Rats were the main culprit in transmitting plague across Europe from the fourteenth to the eighteenth centuries.

and nineteenth centuries by European colonists. Then domestic breeds escaped and became feral horses again.

The largest populations are now found in North America and Australasia. Most groups are managed, particularly to reduce competition for food and space with other wildlife and domestic stock, or to limit their impact on plant diversity. Some populations are unmodified, and others are periodically hunted to control population size.

10. **BULLFROG**

Introduced bullfrogs *(Rana catesbeiana)* are driving North American native frogs to extinction in some areas. Adult males are very aggressive and defend their territories, which can range from 3 to 25 meters of shoreline, by physically wrestling with others. Bullfrogs are predators and usually feed on snakes, worms, insects, crustaceans, other frogs, and tadpoles. They are also cannibalistic and will not hesitate to eat their own kind. There have even been a few reported cases of bullfrogs eating bats.

Unusual Reproduction

Caesarean and *artificial insemination* have become common words to us, even though they were once thought unusual. Those techniques have *nothing* on these practices.

1. **MOTHER NATURE'S CENTRAL HEATING**

Australian mallee fowl *(Leipoa ocellata)* incubate their eggs in a compact pit of rotting vegetation, warmed by the heat of decomposition. The male fowl checks the temperature by plunging his beak into the pile and adds or removes material accordingly, keeping the mound at 34°C.

2. **THE BIG O**

Several nights in late spring, Australia's Great Barrier Reef is the location of the "the biggest orgasm on Earth." As billions of coral polyps swim vertically, the bundles rupture, releasing gametes. They broadcast spawn as the flood tide peaks.

3. **BABY FACTORY**

Also known for carrying her children around long past the comfortable stage, the female kangaroo practices embryonic diapause.

As soon as her young is born and crawls its way from the uterus into her pouch, the female kangaroo breeds again. Fortunately for her and both joeys, the development of the embryo stops after the 100-cell stage. The embryo restarts developing only when its mother can care for it, after the previous offspring stops nursing.

4. A WOLF IN FISH'S CLOTHING

The Amazon molly *(Poecilia formosa)* reproduces asexually by parthenogenesis—the development of unfertilized eggs carrying only maternal genes. The species is entirely female, with rare exceptions.

However, the female's eggs need sperm to trigger development, so the female must trick a male of a related, but separate, species to spend his time mating with no hopes of passing his genes. She does this by mimicking the actions and color patterns of the male's female species.

5. DOWN AT THE LOCAL BAR

The Swedes named the fixed site where male and female animals congregate to find mates the "lek." Males display en masse while defending a small area from the others. Females peruse the selection of posturing males to find an attractive partner, then leave with him to mate. Certain sites can last many years. Animals that use such tactics include the European shorebird called the "ruff"; other birds such as the sage grouse; and some mammals, fish, and insects, such as the cicada-killing wasp.

6. THE DANGEROUS LIFE AS A BACHELOR

Species of gerbils in the Sahara Desert, rodents similar to pet gerbils from Mongolia, live a solitary life. Females inhabit places where adequate food is available for them and their pups.

Males choose habitats closest to females so that they can make their mating rounds. This means that males often inhabit centrally located burrows poor in food reserves and locations that place them in peril for running hundreds of yards in the open.

7. BEYOND PEDOPHILIA

Arizona State University zoologist John Alcock's studies of solitary bees show insect colonies whose female bee mates only once in her lifetime. If the male wants to further his genes, his goal is to be first. This results in males using extreme tactics, such as digging the young virgin females out of their underground abodes before they emerge themselves.

8. DO-IT-YOURSELF MATING

American salamanders place a spermatophore on the ground for the female to sit on and inseminate herself. Often, competing males lay their spermatophores on top of those of their rivals.

9. WINE AND DINE

Additional food enables some female insects to produce more eggs or nurture their young. Using this concept, cricket and grasshopper males will attach a large spermatophore externally to female's genitalia during copulation. To mate again, the female must eat this "wrapper" to remove it.

Some crickets' spermatophore is enormous, sometimes one-third of the male's body weight. Males can take several days to produce another spermatophore, making them extremely fussy about their mates. In this case, it is the females who must compete for male attention.

10. KEVORKIAN SPIDER

The male Australian Redback spider *(Latrodectus hasselti)* goes to great lengths to have sex. When the male begins to inseminate the female, he springs into a summersault and lands on the female's mouth, which forces her to eat her mate. Experimenters note that the action is not nutritionally motivated, because the male is only a fraction of the female's weight, and given the opportunity in nonmating times, the female will not touch the male. However, scientists noted that copulations last twice as long using cannibalism, and females are less eager to copulate with others quickly.

Poisonous Plants

Familiar plants can often be lethal. One example that turned out not to be harmful was the tomato, which had been associated with the poisonous members of the nightshade family. One day while strolling through Lynchburg, Virginia, Thomas Jefferson saw the love apple plant, as it was called then. Jefferson asked Mrs. Owen, the owner of this particular plant, for permission to eat one of the fruits. Much to everyone's amazement, he lived through an entire meal. Lingering doubts about the tomato's safety were supposedly put to rest in 1820, when Colonel Robert Gibbon Johnson announced that at noon on September 26, he would eat a bushel of tomatoes in front of the Boston courthouse. The story goes that thousands of eager spectators turned out to watch the poor man die after eating the poisonous fruits but were shocked when he lived. The plants that follow, however, have always proven to be dangerous.

1. *PAXILLUS INVOLUTUS*

In 1944, Julius Schaeffer, a well-known German mycologist, was found dead in the woods by a colleague, a pile of mushrooms by his side. Destitute during World War II, Schaeffer often disappeared into the woods for long periods to study and live off nature. Further investigation pointed to a large brown mushroom named *Paxillus involutus,* Latin for "turned in." The consumed mushroom produces antigens that stimulate the production of antibodies in the human body, until one day the mushroom finally breaks down the immune system, bursting red corpuscles, spreading poison, and resulting in a cumulative death.

2. HEMLOCK

Hemlock *(Conium maculatum)* is most famous for causing the death of Socrates (c. 470–399 B.C.), who was accused of corrupting Athenian youth and interfering with the religion of the Athenians, and sentenced to death. His disciple Plato described his death as a calm and yet an anxious one: "His legs became colder and colder, he lay upon the garden-seat and kept talking to the people around. He kept his thoughts and mind clear till the very end."

3. JIMSONWEED

All parts—the white, funnel-shaped blossoms, dark green leaves, and the prickly fruit—of the annual jimsonweed are poisonous. In the Middle Ages, it was very popular among professional murderers, who added parts of the plant to victims' food or wine and immediately poisoned them. The affected person develops symptoms similar to those caused by belladonna because of the identical chemical components in both herbs. After eating the plants, sheep have been observed to experience abnormal leg movements, disturbed vision, intense thirst, and biting at imaginary objects in the air.

4. FOXGLOVE

A perennial herb with alternate, toothed, hairy, basal leaves, foxglove's characteristic purple or white pendant flowers are dotted with conspicuous spots on the inside bottom surface of the tube. Several cardiac glycosides—digoxin, digitoxin, and digitonin—contribute to medicinally useful steroids and cortical hormones, but in minute doses the substances are highly poisonous. The glycoside content varies with the individual plant, the stage of development, and the time and method of harvest.

5. OLEANDER

This widely cultivated ornamental evergreen shrub contains a dangerously poisonous juice that contains the toxins oleandrin and nerioside, which are very similar to the toxins in foxgloves. It is a tropical plant but is grown as an ornamental and as a houseplant. Although the plant isn't palatable, it will be eaten by hungry animals. Approximately one-quarter pound of leaves (about 30 or 40) could deliver a lethal dose to an adult horse.

6. CASTOR-OIL PLANT

The only species of the genus Ricinus of the spurge family, the castor-oil plant yields showy foliage, but the seeds grown in the spiny pods are extremely poisonous. The phytotoxin (plant toxin) is ricin, a water-soluble protein concentrated in the seed. Also present are the alkaloid ricinine and an irritant oil. Note, however, that commercially prepared castor oil contains none of the toxin. The seed is toxic only if the outer shell is broken or chewed open. Seeds swallowed intact usually pass without incident. Signs of toxicity may not manifest for 18 to 24 hours after ingestion.

7. YEW

No tree is more associated with the history and legends of Great Britain than the yew. Before Christianity was introduced, it was a sacred tree favored by the Druids, who built their temples near these trees, a custom followed by the early Christians. The two most poisonous parts of the plant are the so-called fruit—bright red, sometimes yellow, and juicy—and the seed. The culprit is an alkaloid, taxine, a poisonous, white, crystalline powder, only slightly soluble in water.

8. BELLADONNA

The poisonous perennial plant *Atropa belladona,* of the nightshade family, is native to Europe and now grown in the United States. The plant has reddish, bell-shaped flowers and shiny black berries. Extracts of its leaves and fleshy roots dilate the pupils of the eyes and were once used cosmetically by women to achieve this effect. The plant extract contains the alkaloids atropine, scopolamine, and hyoscyamine. Belladonna has also been used since ancient times as a poison and a sedative; in medieval Europe large doses were used by witchcraft and devil-worship cults to produce hallucinogenic effects. The active substances physiologically depress the parasympathetic nervous system.

9. WHITE SNAKEROOT

Abraham Lincoln's mother, Nancy, passed away on October 5, 1818, from "milk sickness," a disease contracted by drinking milk from cows that have grazed on poisonous white snakeroot *(Eupatorium rugosum)*.

White snakeroot is a shade-loving plant found throughout Kentucky, Indiana, western Ohio, and Illinois. It grows in the rich, moist soil of woods, thickets, and woodland borders. Milk sickness was most common in dry years when cattle wandered from poor pasture to wooded areas in search of food. As woodlands were cleared by the pioneers, cattle were provided adequate pasture, thus diminishing the incidence of the illness. Because farming techniques have improved, and because the modern dairy industry

10. HENBANE

Henbane *(Hyoscyamus niger)* is a poisonous plant well known since the remote past. It was used in ancient Babylon, Egypt, Persia, Greece, and Rome. It is a biennial grayish-green sticky plant that has an unpleasant smell, and its flowers produce a huge amount of seeds. Henbane contains the same alkaloids as belladonna but is less toxic because the amount is 10 times less. The poisoning effect of the plant rarely leads to death. In most cases, it causes a clinical condition characterized by insanity, violence, seizures, trembling limbs, and other symptoms similar to those caused by belladonna.

In the Middle Ages, henbane was widely used in Germany to augment the inebriating qualities of beer. The names of many German towns originate from the word *Bilsen,* or henbane. Later, the word was transformed to *Pilsen,* to name the famous Pilsen beer. It took many years to prohibit the use of henbane in brewing after numerous cases of poisoning.

Unusual Forms of Life

Just when we think we know every part of the globe, a new discovery tells us to keep exploring. These 10 are just a few of the recent discoveries by scientists who knew there is more in our backyards than grass and ants.

1. LAST FRONTIER

The central ice sheet in Eastern Antarctica never melts, reports *The Economist*. Four kilometers below the frozen surface is Lake Vostok, a freshwater lake as big as Lake Ontario. The average length of time a water molecule stays in the lake is 50,000 years.

Ice samples taken from above the lake contain bacteria, algae, diatoms, and other microfungi. As the lid of ice melts into the lake at 1 millimeter per year, organisms are released into the water. At the point of release, they will be 500,000 years older than when they first landed on the Antarctic icecap.

2. LIFE IN A ROCK

"Endolithic" communities also inhabit Antarctica, *The Economist* reports. In the east polar plateau, cold dry winds averaging 70 mph expose quartz crystals of sandstone rocks, keeping them free of snow. Within rocks, layered communities of black lichens, white fungi, and green algae have been living for centuries.

The creatures can photosynthesize through the little light that passes through the first millimeters of the rock, which stays at an average temperature of 0° and contains some water. The black lichens are the first layer, then the fungi, then the algae.

3. CAVERNOUS LIFE UNDERGROUND

Cristian Lascu discovered the Movile Cave, underneath a cornfield in southern Romania, a few minutes' drive from the Black Sea. The cave was entered for the first time in 1986, after 5.5 million years void of contact. Completely dark and silent, the cave is low in oxygen and high in carbon dioxide and hydrogen sulfide, and it is the first terrestrial ecosystem that does not depend on sunshine. So far, 48 definably different animal species have been found there, 30 on land and 18 in the water. Of these, 33 are particular to the cave. There are no plants or vertebrates, and most animals are troglomorphic—adapted to cave life, with long legs and tentacles, and tiny or nonexistent eyes. Currently found are wood lice, millipedes, bacteria, fungi, pseudoscorpions, centipedes, spiders, leeches, worms, shrimp, and water scorpions. The water contains ingredients for the energy of life—hydrogen sulfide, methane, and ammonium ions.

4. INFRARED EYEBALLS

In 1996, a new species of shrimp was collected from the Snake Pit vent on the mid-Atlantic ridge. It has small eyespots with no lens—instead, the eyes are covered with a smooth sheet of material that is sensitive to light. It is believed to be a type of heat sensor that enables the shrimp to see bacteria living near the undersea thermal vents.

5. SMALLEST LIFE

In 1999, Dr. Philippa Uwins discovered the smallest group of organisms in the world growing in sandstone off the shore of Western Australia, which she named nanobes. The cells are about 50% smaller than what was known before, in a size range that's argued to be too small to exist. The nanobes are also in the same size range as the Martian nanobe bacteria that were found in a meteorite some years ago. Morphologically they look similar to fungi and may be similar to Archeae in that they can withstand extreme temperatures.

6. FLOATING AMAZON

The Sargasso Sea is a free-floating (pelagic) kelp ecosystem in the western North Atlantic. Two species constitute the algae there: *Sargassum natans* and *Sargassum fluitans.* These kelps provide a floating home for animals that don't swim very well, such as crabs, so that if they were to fall, they would sink to the depths below.

Many organisms are omnivores, not limiting themselves to one food source. Some of the more unusual forms include fish and crabs that are camouflaged to look like Sargassum. An example is the pipefish *Syngnathus pelagicus,* a brownish-green relative of the seahorse that is covered by flaps of skin.

7. A PLANT THAT NEEDS ALGAE

Cycads are unique among gymnosperms because they are the last example of the older era of plants, somewhere between ferns and conifers. They have a remarkable nitrogen-fixing cyanobacteria, also called blue-green algae, living inside root structures in the plant's base.

Like bacteria in clover and bean plants, cyanobacteria supply the cycad with nitrogen through its own biological processes. In return, the cycad supplies the cyanobacteria with food. Although the cyanobacteria are below ground, they still possess organs for photosynthesis.

8. GIANT BACTERIA

The biggest bacterium is a single-celled ocean dweller named *Thiomargarita namibiensis,* which means "sulfur pearl of Namibia." It was found in the ocean floor off the coast of Namibia, Africa. Its ball-shaped cells can grow to almost 1 millimeter and can be seen with the unaided eye.

The bacterium consumes sulfur and releases nitrate, which it stores in vacuoles reported to take up 97% of the cell. They give the bacterium a pearly, blue-green color.

9. 3.5-MILE ORGANISM

The honey mushroom *(Armillaria ostoyae)* is currently seen as the largest organism, measuring 3.5 miles in diameter and aged approximately 2,400 years, mostly hidden underground.

Feeding on trees, the fungus was discovered when Catherine Parks, a scientist at the Pacific Northwest Research Station in La Grande, Oregon, heard about a big tree die-off in 1998. Parks used aerial photographs and collected root samples to identify the fungus through DNA testing, finding that a single fungus had grown to gigantic proportions.

10. ARCHAEA, OR "ANCIENT LIFE"

Methanococcus jannaschii thrives at boiling temperatures, eats iron and sulfur, and expels natural gas. The organisms live about 2 miles beneath the Pacific Ocean. DNA analysis shows that it is not a bacterium, but possibly a form of life called Archaea, dating back billions of years. Two-thirds of its genes look brand new, says J. Craig Venter, president of the Institute for Genomic Research, as reported by *Popular Mechanics* writer Jim Wilson.

Microlivestock

Entomophagy (the eating of insects) has yet to become a favored practice in the United States and Europe, in spite of the superior nutritional content of edible insects compared with other animals. But even in these cultures, humans unknowingly consume 1 to 2 pounds of insects each year. Invisible to the eye, they have been ground up into tiny pieces in such items as strawberry jams, peanut butter, spaghetti sauce, applesauce, and frozen chopped broccoli. These insect parts make some food products more nutritious. Other cultures have made insects a main ingredient in their diets, the top 10 listed here.

1. ANTS

The most popular group of edible insects is also the most globally widespread. One example, the honeypot ant in Australia, is a favorite sweet treat of the Aborigines. The species' worker ants gather honeydew and pass it along to other workers, which head for the "larder" ants. These living storage bins keep the nectar in their bellies until requested to give some up to workers. Foragers sometimes must dig as deep as 2 meters to find the subterranean nests, but when they do, the heads of the ants are pinched off and the remainder eaten raw.

2. **GRASSHOPPERS**

Next to ants, grasshoppers, or locusts, are the second-most popular insect food. Asians consume locusts, considered the most destructive insect in the world, as a snack rather than a dish, usually deep-fried. Africans eat them raw, fried, roasted, jellied, or smashed into paste.

3. **SPIDERS**

Indians in South America, bushmen of South Africa, and the Aborigines eat spiders commonly. The Chinese also add medicine to the list of arachnid uses. Live spiders are rolled in butter and swallowed whole, or eaten in molasses, or rolled in the cobweb and taken like a pill.

4. **WATER BUGS**

Water bugs, one of the largest insects, are eaten with a fish sauce after being roasted over a fire, steamed, pounded, or added to chili paste. The critters reach 3 inches in length, but connoisseurs remove the carapace, wings, and legs before consuming.

5. **DUNG BEETLES**

Dung beetles are soaked in a bucket of water overnight to allow them to rid themselves of ingested matter. They then are soaked in another bucket of clean water for two to three hours, and are fried in a covered pan without oil.

6. **WEEVIL**

In West Africa, coconut larvae is a dish offered only to good friends. Coconuts at the half-hard stage are emptied of milk, refilled with larvae, recapped, and then boiled.

7. CATERPILLARS

The 1987 Thai Ministry of Public Health included silkworm pupae on a list of local foods that could be used in supplementary food formulas for malnourished infants and preschool children. The spiny mopane caterpillar is so popular that when it is in season, the sale of beef declines. South African government researchers claim that 20 of the caterpillars will satisfy an adult male's daily requirement for calcium, phosphorus, riboflavin, and iron.

8. BUSH TUCKER

Aborigines collect moth grubs by digging up the roots of the acacia bush to obtain grubs, which are eaten raw or in any cooked manner. Dr. Ron Cherry, author of *Cultural Entomology Digest*, says only 10 grubs are sufficient for the daily protein needs of an adult.

9. DRAGONFLIES

Pepes is a Balinese dish in which dewinged and delegged dragonflies are added to a pounded mixture of coconut paste, fermented fish paste, garlic, chilies, tamarind juice, basil leaves, ginger, and lime. The entire mixture is wrapped in banana leaf packets and roasted.

Most dragonflies are caught with nets, but some hunters observed the dragonflies landing on the ends of reeds and fishing lines and thus devised a sticky pole to catch them.

10. TERMITES

Called bushman's rice or rice ants, termites are the third-most consumed insects, largely in Africa. Natives capture them when they swarm or attract them by placing a bowl of water under a light source at night. High in calories and a rich source of fat, the oil from their bodies is collected and used later to flavor dishes with a buttery taste. African termite queens may reach the size of a large potato, and each can lay 100 million eggs in its lifetime.

Weather-Forecasting Tools

For centuries, we have tried to predict the weather—and have never truly succeeded. Our drive to capture the elusive whims of nature has forced us to turn to nature's signals.

1. PLANT OIL

Water vapor has no odor, but approaching rain can be smelled. Plants ooze oils into the soil. When air pressure drops, the soil releases oils, producing the "smell" of rain. Additionally, dropping temperatures and moisture in the air amplify smells.

2. OPTICAL ILLUSION

The optical illusion that faraway objects appear close before a storm is caused by a difference in temperatures between levels of air. When upper air is warmer than normal, the light is refracted differently, disrupting the view. The mirage is called the Hillenger effect.

3. BAROMETRIC PRESSURE

Some people feel rain in their bones, especially if they suffer from arthritis or rheumatism. Numerous studies have found that a drop in barometric pressure brings discomfort to those who suffer joint diseases.

4. CHURCH BELLS

Sound travels better in high humidity. City dwellers often gauge the chance of rain by the clarity of church bells ringing in the distance. "Sound traveling far and wide, a stormy day will betide."

5. ACTIVE INSECTS

When air pressure drops, insects become more active.

6. CICADAS

In dry weather, cicadas hum loudly, using their stiff wings, but in high humidity, their wings soften, causing them to hum lower.

7. SPIDERS

"When spider webs in the air do fly, the spell will soon be very dry." Spider webs absorb moisture from the air. When the air is dry, they can be easily pulled from their perches.

8. COTTONWOODS

Cottonwoods turn up their leaves before a rain.

9. DAISIES

Daisies close up before a rain.

10. SWALLOWS

"Swallows fly high, clear blue sky. Swallows fly low, rain we shall know." Swallows follow the insects that they eat, which fly lower toward the ground before a rain.

Worst Extinctions

Creatures on Earth today represent a tiny fraction of the total species that have existed on this planet. Global warming has scientists hot and cold over worries that the human race is imperiled. Some believe it should have top billing on our government's spending list, whereas others believe it's just hype. If fossil records mean anything, there is no doubt that we should be stocking up on heat sources. The key questions are, When? and Will the warming trend kill off the current life on Earth?

1. GLOBAL WARMING AND HUNGRY HUMANS

About 13,000 years ago, a global warming trend signaled the end of the last ice age. Forests invaded the plains, and human predators helped push many grazing species into extinction (known as the Pleistocene Megafauna kill). Hunted animals included the wooly mammoth, wooly rhinoceros, steppe bison, giant elk, European wild ass, and an entire genus of goats. Horse and cattle numbers dwindled in Europe. Scientists estimate that a total of 32 large animal genera were killed off (or hunted to extinction).

2. HUMANS 1, DINOSAURS 0

The extinction that marked the end of the Cretaceous period was not the largest, but it wiped out dinosaurs and gave mammals a chance to thrive. However, some argue that dinosaurs didn't die out in the end of the Cretaceous period—rather, they had formed wings in the previous Jurassic period and flew away.

3. A FAUNA BEFORE TIME

R. C. Sprigg, a mining geologist for the Australian government, explored old lead mines for a living, but luckily for paleontologists, he had a strong amateur interest in fossil collecting. He kept his eyes peeled for rocks no normal paleontologist would see. In 1946, Sprigg (and Sir Douglas Mawson) discovered the Ediacara fauna, one of the oldest forms of life, in pure quartz, which was not known to contain fossils. The aquatic, soft-bodied "jellyfish" showed that 600 million years ago, a major evolutionary branch had died. Further studies showed that the fossils were found well below the oldest Cambrian strata, actually dating from the Precambrian era.

4. COLD SETS IN

In the Precambrian era, about 650 million years ago, prokaryotes practiced primitive living as single cells. However, eukaryotic Acritarch algae (more than one cell) were found extinct in fossils in Scandinavia, Africa, and Australia. This era is regarded as the greatest episode of continental glaciation.

5. GREAT AUNT HORSESHOE CRAB REMOVED

The Cambrian era (570–504 million years ago) saw the first hard-shelled animals. The trilobite arthropod looked like a horseshoe crab with more segments. Archaeocyathids were cone-and vaselike reef builders, while Brachiopods, often called lamp shells were also apparent. Conodonts were eel-shaped swimmers recognized in fossils by their tooth-like structure. This era saw the beginning of a glaciation.

6. NAUTILUSES LEFT BEHIND

The Ordovician era (440–450 million years ago) saw a broad expansion of life left from the small players in the Cambrian era, but like others, came to an end. With it ended the reign of the nautiloid. Related to squid and octopus, modern nautiluses are sole survivors. Early species were small and straight, but some grew up to 10 feet. Sea lilies, rugose corals, and cup corals that did not form colonies all proliferated. Marine life, invertebrates, and reef builders all showed mass extinction.

7. REEF COMMUNITY CLEARED

In the Devonian era (360–370 million years ago), armored jawfish reached up to 30 feet. The era saw the decimation of the reef community that had lasted through the previous Ordovician period's extinctions. Reefs are not thought to be well established until hundreds of millions of years later, but silica sponges thrived. Higher plants, terrestrials, tall as trees, were not as affected as the marine fish, plankton, marine invertebrates, and primitive fish.

8. THE GREAT EXTINCTION

In the largest extinctions of all, at the end of the Permian era (250–255 million years ago), about half of all existing families disappeared, including vertebrate land animals. One history-altering loss was Therapsids, mammal-like reptiles possibly partially endothermic (warm-blooded). Of all preexisting species, 75 to 90% went extinct, including seafloor protozoans and reef builders.

9. MAMMAL-LIKE REPTILES LOST

The Triassic period began with a rapid growth of mollusks. Next came the "Sea Monsters," or early dinosaurs, then turtles and crocodilians, frogs, and small mammals.

This period is the most mysterious, with fossil evidence showing that land extinctions occurred before sea extinctions. The end of the era revealed the death of most mammal-like reptiles. Fewer than 10% of Triassics lived into the next era, and only 20% of pre-existing families survived. Marine invertebrates died.

10. PLANT POPULATIONS CHANGE

The end of the Cretaceous period showed mollusks and planktonic foraminiferans (ocean floaters) dying off. Also, the nannoplankton deposits that now give us chalk stopped. Cases of fern spores show that sporangial plants dominated for a short period of time, and then pollen-using plants reestablished themselves. Species of dinosaurs, marine reptiles, and reef builders died.

PALEOBIOLOGY

Dinosaur Discoverers

Dinosaurs appeared long before humans, the latest evidence pointing to approximately 230 million years ago in the Triassic period. However, it wasn't until the eighteenth century that we began to identify dinosaur fossils, and we have yet to piece together a definitive statement about their lives and times.

The first dinosaur bone was discovered by Robert Plot, professor of "Chymistry" at Oxford. His weighty specimen was found in a shallow limestone quarry at Cornwall in Oxfordshire. He postulated that it might have come from an elephant brought to Britain by the Romans or it was a mythical giant. Plot's bone was later identified as a Megalosaurus, though in 1763 illustrator R. Brooks labeled it *Scrotum humanum,* in honor of its appearance.

1. DR. GIDEON MANTELL

As a surgeon, Dr. Gideon Mantell (1790–1852) studied geology and fossils as a hobby. In 1822, while Mantell visited a patient in the Cuckfield region of Sussex, the story goes that his wife, Mary Ann, who shared his passion for fossils, found fossilized teeth in a pile of stone intended for mending roads. Mantell was intrigued by the largest tooth, which seemed to belong to a plant-eating mammal, an anomaly since at that time no mammals were known to have existed in Cretaceous rocks. Others who examined the tooth said that it belonged to a fish or rhinoceros. In 1824, another naturalist showed Mantell an iguana skeleton, and he was struck by a remarkable similarity. Mantell planned to call the animal Iguanosaurus, but the Rev. William Conybeare suggested Iguanodon (iguana tooth), and Mantell agreed. Mantell also named the Hylaeosaurus, the Pelorosaurus, and the Regnosaurus.

2. SIR RICHARD OWEN

Sir Richard Owen (1804–1892) originally defined the clade Dinosauria on August 2, 1841, in a two-hour speech on British fossil reptiles at a conference of the British Association for the Advancement of Science in Plymouth. He explained that the term *Dinosauria* meant "fearfully great lizard." At 32 years of age, in 1836, he had become Hunterian professor of the Royal College of Surgeons, with access to the museum in which Mantell had made the connection between his tooth and the modern iguana.

Owen believed that dinosaurs had pillar-like legs, not angled ones characteristic of sprawling reptiles. His motive for describing them this way may have been his staunch antievolutionary stand, since it appeared to him that these superior creatures had left no highly evolved descendants.

3. EDWARD DRINKER COPE

The eldest child of a Quaker family of merchants, Edward Drinker Cope (1840–1897) began drawing fossils in the Museum of the Academy of Natural Sciences at the age of 8 and published his first scientific paper at 18. As a teen Cope had been sent by his father to work on a farm, but he persuaded his father that lectures on anatomy could help him work with farm animals. He joined the Academy of Natural Sciences and recatalogued its reptile and amphibian collection. Cope's drive led him to become a tremendous

Sir Richard Owen, the man who originally defined the clade Dinosauria in 1841. He explained that the term *Dinosauria* meant "fearfully great lizard."

force in vertebrate paleontology, especially that of the American West, and also to butt heads with another strong-willed paleontologist, O. C. Marsh, in one of the most noted of scientific feuds. Cope named more than 1,000 species from fossils.

4. O. C. MARSH

O. C. Marsh's (1831–1899) uncle was financier George Peabody. Because his mother died when he was age 3, he spent a lonely childhood collecting fossils as the Erie Canal was widened. Supported by his uncle, he studied at Yale and later talked his uncle into a $150,000 endowment for a museum at Yale, which brought Marsh a professorship. Throughout his career, Marsh described and named approximately 500 new species.

At first on friendly terms with Edward Drinker Cope, even naming fossils after one another, Marsh ridiculed Cope's erroneous restoration of a marine reptile skeleton. Between 1870 and 1873, Marsh led four expeditions of Yale students into the West. Cope traveled there in 1872 and beat Marsh to the description of a Cretaceous horned dinosaur. Marsh felt that Cope had trespassed on his turf, so he convinced a member of Cope's party to spy for him. When Cope tried to telegraph news of his discovery, Marsh bribed the telegrapher to give him the information.

5. FERDINAND VANDEVEER HAYDEN

Surveyor Ferdinand Vandeveer Hayden (1829–1887) is credited with the first discovery of dinosaur remains in North America in 1854 at the confluence of the Judith and Missouri Rivers. After Lewis and Clark's expedition of 1804–1806, a number of treks were undertaken to make geological surveys. Hayden became well known to the Sioux, who named him "Man Who Picks Up Stones Running" and regarded him as a lunatic for collecting fossils. The Blackfeet, however, drove him out of their territory. In 1855, Hayden brought back fossil teeth to the Academy of Natural Sciences in Philadelphia, where paleontologist Joseph Leidy described them as belonging to Deinodon, or "terror tooth," and Trachodon, or "rugged tooth."

6. BARNUM BROWN

American paleontologist Barnum Brown (1873–1963), Henry Fairchild Osborn's (1857–1935) favorite dinosaur hunter, was born a few days before Phineas T. Barnum's "Great Traveling World's Fair" arrived in Carbondale, Kansas. When Brown attended the University of Kansas, he studied under O. C. Marsh's former assistant, Samuel Wendell Williston. On a university field trip to Wyoming in 1895, Brown found his first dinosaur fossil, a Triceratops skull. Osborn commented later, "He must be able to smell fossils. If he runs a testtrench through an exposure, it will be right in the middle of the richest deposit. He never misses."

In Montana in 1902 and 1908, Brown found the Tyrannosaurus skeletons now adorning museums in Pittsburgh and New York City. While collecting for the American Museum of Natural History in 1910, he started the "dinosaur rush" along Alberta's Red Deer River. He conducted a friendly rivalry with C. H. Sternberg of the Geological Survey of Canada, who became the leading freelance collector of his day after the deaths of Edward Drinker Cope and Marsh.

7. ROY CHAPMAN ANDREWS

After World War I, the American Museum funded an expedition to the Gobi desert to search for the missing link, and naturalist-explorer Roy Chapman Andrews (1884–1960) led the trek. The story goes that Andrews, with only $30 in his pocket, had asked the American Museum director if he could scrub floors in order to learn more from the museum's collection but instead was hired to help set up exhibits. He soon became an expert on cetaceans. When Andrews heard Henry Fairchild Osborn's theory that Asia was the birthplace of humanity, he led five major Central Asiatic expeditions between 1922 and 1930. Red Gobi sandstone preserved innumerable skeletons, among them Andrews's own horned dinosaur, *Protoceratops andrewsi*. In 1925, the largest of all expeditions at the time set out from Peking with 40 men and their vehicles and 125 camels.

In 1923, Andrews returned to the United States with Iguanodon eggs and was besieged with offers of thousands of dollars for the exclusive right to photograph them. An egg auction and other fundraisers earned $280,000 to prepare for another expedition. When Andrews gave his lecture at the American Museum, 4,000 people fought for 1,400 seats. The character of Indiana Jones was based partly on the dashing explorer, who always bore his rifle.

8. FRIEDRICH VON HUENE

It wasn't until the twentieth century that Germany's larger dinosaurs received much attention, but the country's chief researcher appeared in 1875, shortly before the discovery of the Berlin Archaeopteryx. Friedrich von Huene (1875–1969), while studying at the University of Tubingen, became a specialist in the earliest dinosaurs of the Triassic period. Around 1900, he began work in the valley of the Necker River, where Upper Triassic rocks were well exposed and, in 1908, published a monograph in which he discussed the Plateosaurus. After World War I, von Huene opened a dinosaur quarry on a hillside in a wooded valley near Trossingen, where he found thousands of bones belonging to Plateosaurus. He continued to specialize in Triassic dinosaurs and worked with typical energy into his old age. In his late eighties, von Huene astonished a colleague by taking a 100-mile, three-day hike to a scientific meeting.

9. DONG ZHIMING

Chinese paleontologist Dong Zhiming named 19 Chinese dinosaurs between 1973 and the late 1980s. During the Cultural Revolution, some paleontologists were sent to farms to work or were placed under house arrest. Despite these obstacles, a new generation of paleontologists was trained. In 1979, when an industrial plant was under construction at Dashampu, Zhiming was called in to develop the rich Middle Jurassic quarry, which provided more than 100 dinosaur skeletons. Zhiming honored writer-producer Michael Crichton (author of *Jurassic Park*) with his own dinosaur, Crichton's ankylosaur, which resembles the Scutellosaurus.

10. ROBERT T. BAKKER

Vertebrate paleontologist Robert T. Bakker (b. 1946) served as a model for the scientist in Michael Crichton's *Jurassic Park*. His trademark white Stetson hat and long beard fit his image as an outspoken bone hunter. Bakker earned his doctorate from Harvard and is the curator at the Tate Museum in Casper, Wyoming. In his book, *The Dinosaur Heresies,* he discredits the meteorite theory of extinction and suggests that viruses caused the dinosaurs' demise. Also, Bakker feels he has proved that dinosaurs are warm blooded and share ancestry not with reptiles but with birds.

Recently Discovered Dinosaurs

The clade Dinosauria was originally defined by Sir Richard Owen in 1842. In the 1990s and the early twenty-first century, a burst of dinosaur discoveries has reshaped paleontologists' perceptions of the animal. Recent discoveries highlight dinosaurs' international scope of interest as well as their diversity. Dinosaur aficionado George Olshevsky, or "DinoGeorge," of San Diego, California, has catalogued more than 600 valid kinds of dinosaurs.

1. CARNOTAURUS

Argentinian paleontologist José Bonaparte was the first to excavate giant dinosaurs in South America. In 1985, he discovered several theropods in Patagonia, a treeless plateau in southern Argentinia. One was a 1-ton behemoth with horns above its eyes. Therefore, he named it Carnotaurus, or meat-eating bull. It also had an unusually large braincase.

2. CRYLOPHOSAURUS

From 1990 to 1991, Dr. William Hammer of Augustana College in Illinois unearthed Crylophosaurus, or frozen-crested reptile, on the slopes of Mount Kirkpatrick in the Antarctic. Probably dating from the early Jurassic period, the theropod's fossil proves that dinosaurs lived on all continents and in high-latitude climates. Ferns, evidence of a temperate climate, were found alongside the 200-million-year-old dinosaur. The predator probably killed a herbivore and choked to death while eating the flesh. Other predators then gnawed on the big killer, and mud later entombed the carcass.

3. ARGENTINOSAURUS

In 1993, José Bonaparte dug up an even larger sauropod in Patagonia. Bonaparte estimated that the herbivore, which he named Argentinosaurus, measured, counting its ultralong neck, approximately 148 feet long and weighed 100 tons, making it the largest creature to have roamed Earth.

4. AFROVENATOR

In 1993, paleontologist Paul Sereno of the University of Chicago unearthed a large, relatively intact skeleton of a Cretaceous carnivore in Niger. Sereno's team named it Afrovenator, or African hunter.

5. GIGANOTOSAURUS

In 1995, paleontologist Rudolfo Coria of the Carmen Funes Museum in Argentina and Leonardo Salgado unearthed a *Tyrannosaurus rex* look-alike that they dubbed Giganotosaurus. It measured 46 feet long and weighed 8 tons. Although the theropod was enormous, its brain was only the size of two soft drink cans placed end to end—half that of a *T. rex*.

6. NOTHRONYCUS

A team led by Doug Wolfe of the Mesa Southwest Museum in Arizona discovered a 1-ton, potbellied vegetarian that they named Nothronycus. Paleontologists picture it as a 2-foot-tall feather-coated biped with a tiny head and long neck. Each of the neck vertebrae is bigger than its skull. Huge claws protruded from its forelimbs.

7. PROTARCHAEOPTERYX ROBUSTA

The early Cretaceous site in Liaoning, China, yielded well-preserved fossils of soft tissue, including feathers and down of the *Protarchaeopteryx robusta,* or robust first ancient winged one. The 3-foot-long theropod isn't in the class Aves. Unlike flying birds, it wasn't capable of sustained powered flight and had symmetrical feathers (without the narrow leading edge needed for true flight). Feathers covered its body and formed "wings" on its forelimbs and large fans on its tail. Some scientists say that this discovery supports the theory that dinosaurs gave rise to birds.

8. SAUROPOSEIDON

In 1999, a team of scientists from the University of Oklahoma announced the discovery of one of the largest dinosaurs that ever lived, Sauroposeidon, or earthquake god lizard. Dr. Richard Cifelli of the Oklahoma Museum of Natural History discovered the sauropod's vertebrae in 1994. The late Jurassic specimen weighed approximately 55 to 60 tons, and its neck measured 37 to 39 feet long.

9. PARALITIAN STROMERI

In 1999, Joshua Smith, a graduate student at the University of Pennsylvania, entered the wrong coordinates into his global positioning system receiver while in the Bahariya Oasis in Egypt. He had placed a bet over several beers with friends that he could find a paleontological gold mine excavating where a Bavarian geologist, Ernst Stromer, had unearthed extensive fossils in the 1930s. Stromer's finds, housed in a Munich museum, were destroyed during Allied bombing in 1944, and no one had excavated the site since. Smith came upon a manuscript in Cairo that contained the latitude and longitude of one of Stromer's quarries. Having entered the wrong coordinates, Smith ended up in the Egyptian desert and quickly got lost. Sticking his head out of his Toyota Land Cruiser to find his way, he spotted a huge bone. It turned out to be that of a titanosaur, a long-necked quadrupedal plant eater that lived during the Cretaceous period. This one was the second-most massive ever discovered. Researchers named it *Paralititan stromeri,* or tidal giant. It probably stretched to 90 to 100 feet in length and weighed 150,000 to 160,000 pounds. Only Argentinosaurus was heavier, but probably not longer. The skeleton lay in what was once a tropical mangrove swamp. The pelvis had been ripped apart; a nearby tooth belonged to the likely diner, a Carcharidontosaurus.

10. **GALLIMIMUS BULLATUS**

In 2000, Dr. Peter J. Makovicky, then a member of an expedition of the American Museum of Natural History in New York City, discovered a fossil in the Gobi desert of Mongolia of an ornithomimid, an ostrichlike creature belonging to the group theropod. The primitive ornithomimids had teeth, but they later evolved a toothless beak. Makovicky's specimen had a comblike plate in its jaw, similar to the filter-feeding structure in a duck's bill, which had never been seen before in a dinosaur. The fast-running ornithomimid, which means bird mimic, may have eaten by straining invertebrates such as tiny shrimp and other food particles from water and sediment.

Why Dinosaurs Became Extinct

Dinosaurs first appeared on the scene 220 million years ago and dominated the land for more than 150 million years. Did they suddenly disappear after a huge meteorite hit Earth, did they become crispy critters during a volcanic eruption, or did they just slowly fade away? The following theories—some serious, some not—are only a few of the ideas mankind has concocted to explain the mystery.

1. **CONSTIPATION**

In 1964, E. Baldwin reasoned that the dinosaurs died of constipation. At the end of the Cretaceous period, plants containing laxative oils became scarce. Therefore, dinosaurs living where the plants no longer existed acquired stopped-up plumbing and died.

2. **POLLEN**

In 1983, sedimentary geologist R. H. Dort Jr., fed up with suffering the effects of pollen, suggested that dinosaurs became extinct because of "itching eyes."

3. **RACIAL SENESCENCE**

An outdated and discarded theory stated that the dinosaur lineage simply grew old and "senile." Thus, late-appearing species were not as robust as early species and were prone to disease.

4. **VOLCANOES**

In 1972, Peter R. Vogt of the Naval Research Laboratory in Washington, D.C., pointed out that extensive volcanism had occurred at roughly the Cretaceous-Tertiary, or K/T boundary, principally in India. The volcanism produced extensive lava flows, known as the Deccan Traps (*decca* means *southern* in Sanskrit, and *trap* means *staircase* in Dutch). Individual lava flows in the Deccan Traps extend well over 3,900 square miles and have a volume exceeding 2,400 cubic miles. Vogt suggested that the traps may be connected to the Cretaceous extinctions.

5. **CARBON DIOXIDE**

In the mid-1970s, Dewey M. McLean of the Virginia Polytechnic Institute proposed that volcanoes could produce mass extinctions by injecting huge amounts of carbon dioxide into the atmosphere that would trigger abrupt climate changes and alter ocean chemistry.

6. **GIANT METEOR IMPACT**

In 1980, Nobel laureate physicist Luis Alvarez, his son Walter, and their colleagues suggested that an asteroid impact wiped out the dinosaurs and many other species 66 million years ago. The impact left its mark as a 180-kilometer hole and also a thin layer of iridium-rich debris deposited in clay throughout the world at the K/T boundary. This layer, found beneath the ocean floor and in other erosion-free zones, marks the juncture between the two geologic eras. It also correlates with the date of mass extinction. Alvarez proposed that the iridium layer might be of extraterrestial origin.

One of the largest impact craters was found in Chicxulub on Mexico's Yucatan peninsula. Geologist Eugene Shoemaker said that the crater represents one of at least two separate impacts that did in the dinosaurs. After the "big one" (roughly the size of Washington, D.C., and weighing a trillion tons) formed Chicxulub, Shoemaker theorized, a smaller body landed in Iowa, leaving a 35-kilometer-wide crater. Others say that it may have landed in the Pacific Ocean. The impacts could have spawned prolonged darkness, global warming, acid rainfall, and even global wildfires.

Some argue that the K/T impact may not have been unique or as devastating as projected. The problem lies in the fact that other impacts did not cause extreme extinctions of dinosaurs. The late Triassic Manicouagan impact in Quebec and the late Jurassic Morokweng impact in South Africa, which may have exceeded the power of Chicxulub, perhaps wiped out a few groups of dinosaurs. The Cenozoic Popizai-Chesapeake impact in Siberia doesn't coincide with a major extinction, either.

7. **CRISPY CRITTERS**

A team of American oceanographers believes that the dinosaurs might have died in a gas-fueled firestorm. If a giant asteroid or comet hit the Gulf of Mexico, it would have generated huge shock waves that would have freed huge quantities of methane trapped in sediment 500 meters below seabeds. At these depths, low temperatures and high pressure cause methane to combine with water to form methane hydrate. If this were tossed into the air, lightning could have ignited it and turned the atmosphere into an inferno.

8. BAD AIR

After having analyzed gas bubbles trapped in amber, geologists claim that the dinosaur disappearance was caused by bad air. Theorists calculated that the amount of oxygen in the atmosphere fell from 35% to 28% over the course of 500,000 years, and that the dinosaurs' lungs couldn't adapt to the change.

NASA scientists have an alternate explanation. They noted that the rocks near Chicxulub contain abundant amounts of sulfur. They believe that a meteor's impact vaporized the sulfur and spewed more than 100 billion tons of it into the atmosphere, where it mixed with moisture to form tiny drops of sulfuric acid. These drops then created a barrier that could have reflected enough sunlight back into space to drop temperatures to near freezing and could have remained airborne for decades.

9. HURRICANES

In 1995, MIT atmospheric scientist Kerry A. Emanuel proposed that monster hurricanes, called hypercanes, could have arisen if ocean water warmed to 50°C, almost twice the current tropical temperature. A large meteorite impact into the ocean or a major volcanic eruption in shallow water could have driven temperatures into the hypercane zone, creating storms that reached altitudes of 45 kilometers, well up to the stratosphere. Dust, particles, and water vapor would thus block sunlight and decimate the ozone layer.

10. SIZE

Theories that the dinosaurs became extinct because they were so large have largely been discredited. Gregory S. Paul, a freelance scientist and artist who contributes to *Scientific American* and is the author of *Predatory Dinosaurs of the World*, explains that animals follow two reproductive types: k-strategists, such as mammals, which reproduce slowly and spend extensive time rearing their young; and r-strategists, which produce large numbers of young and disperse quickly when conditions are right. Paul believes that dinosaurs were r-strategists because few dinosaur groups went extinct before the end of the Mesozoic era, and they diversified, with older groups continuing to live alongside the new. The enormous sauropods were diverse for 130 million years. By contrast, elephants have been extant for only 40 million years.

All in the Family?

The more scientists dig, the more hominid species they find. Most are distant cousins that went extinct; others are our direct ancestors.

1. ***ARDIPITHECUS RAMIDUS***

The earliest group, the species known as Ardipithecus, or ground ape, dates from 4.4 million years ago. It was recovered from the site of Aramis in the Awash River Valley of Ethiopia in 1994. The group appears to have primitive apelike skeletal anatomy, with small molar and large front teeth.

2. **AUSTRALOPITHICUS**

Australopithicus, or southern ape, is a widespread genus of early hominids found in East and South Africa. Fossils date from more than 4 million years to around 2 million years ago. Two species, *A. anamensis* and *A. afarensis,* are the most ancient of the group. Found principally in Southern Africa, a third group called *A. africanus* dates from between 3 million and 2 million years ago. Australopithecines had relatively small brains, large molars and premolars, long upper limbs and short lower ones, and large differences in size between males and females. Yet, in spite of these differences, they resemble humans more than apes because they walked on their two lower limbs. The most famous *A. afarensis* specimen, Louis Leakey's Lucy, was found in Ethiopia in 1974.

3. PARANTHROPUS

Paranthropus, or similar to man, existed at approximately the same time as *A. africanus* and early Homo. The large-toothed, small-brained hominid had robust jaws and teeth, which led scientists to the nickname of "robust" australopithecines. This bipedal hominid spent more time on the ground than its predecessors because it possessed human-like hands and feet. The hominid ate mainly grasses and plants, while Austrolopithecus ate a variety of foods, both plant and animal. The earliest specimen was found in northern Kenya and dates from around 2.5 million years ago. Known as the "black skull" because of its dark staining, the specimen has a small braincase with large crests for the attachment of huge chewing muscles.

4. *HOMO HABILIS*

Homo habilis, or handyman, is best known from the fossils found at Olduvai Gorge in Tanzania, but fossils have also been recovered in Kenya and Ethiopia. *H. habilis* had a larger brain and smaller, narrower premolar teeth than Australopithecus. The larger brain, certain human-like features of the hands and feet, the proportion of upper and lower limbs, and the use of tools indicate that the group were human-like, though they were adept tree climbers.

5. *HOMO RUDOLFENSIS*

Homo rudolfensis dates from the same time period as *Homo habilis,* yet, geographically, it is limited to northern Kenya. The skull, known as ER 1470, was discovered by Richard Leakey and shows the characteristic large brain and broad, flat face and cheekbones of the Homo line. The molar and premolar teeth are broader than those of *H. habilis.*

6. *HOMO ERECTUS*

Homo erectus, or upright man, dates from the Pliocene-Pleistocene epochs, beginning around 1.8 million years ago. Unlike other early hominids, *H. erectus* is found not only in Africa but also in Asia and Europe. *H. erectus* had a larger brain with thick skull bones and prominent brow ridges. The crests on the skull for the attachment of chewing muscles were smaller than those of early hominids, and *H. erectus* appears to have been taller and heavier. The best-known specimen, dated to 1.6 million years ago, is of a 12-year-old boy who was about 6 feet tall. Some say that the boy died of a crippling disease. *H. erectus* sites contain more advanced tools and remains of large mammals, which point to hunting and the use of fire.

7. *KENYANTHROPUS PLATYOPS*

In 1999, Justis Erus, of the famous Leakey "hominid gang," unearthed a skull of a new prehuman species, *Kenyanthropus platyops,* at a site called Lomekwi in northern Kenya. Dating from between 3.5 million and 3.2 million years ago, the fossil falls in the same time period as Australopithecus, making it a coexisting species. *K. platyops* had a flat face and small teeth, from which scientists conclude that it probably ate fruit, berries, and insects.

8. *ARDIPITHECUS RAMIDUS KADABBA*

Yohannes Haile-Selassie of the University of California at Berkeley discovered the bones of a human ancestor who may have killed animals for food. The key fossil was a partial skeleton of a small hominid from the genus Ardipithecus, as well as antelope bone fragments bearing telltale marks left by stone implements. Believed to have lived 2.5 million years ago, its braincase, face, and palate were more primitive than Homo's, though the face projected forward. The braincase was crested and small. The team discovered a total of 11 specimens from at least five different individuals.

9. *HOMO NEANDERTHALIS*

In 1856, workers digging in a quarry in Germany's Neander Valley unearthed an aged skeleton. Its stooping limbs and heavy brow led anthropologists to conclude that it was the remains of a diseased Mongolian Cossack who had died during the Napoleonic Wars. But subsequent discoveries of similar skeletons eventually revealed that the fossils belonged to a human-like creature who had lived thousands of years before. Dubbed Neanderthal, the ancient being not only gave scientists the first clear evidence for Darwin's suggestion that humans had descended from an apelike ancestor, but its hulking form also etched the image of the brutish caveman indelibly in the public mind.

More than a century later, new findings continue to redefine humanity. Neanderthals were intelligent creatures who were adapted to their Ice Age environment. By dating bits of flint that had been discovered along with a Neanderthal skeleton, researchers found that the Neanderthal lived as recently as 36,000 years ago, at least several thousand years after modern humans are thought to have arrived in Europe. Anthropologist Christopher Stringer of the Natural History Museum in London, one of the proponents of the Out of Africa Model, says that Neanderthals were an evolutionary "dead end" that disappeared with the arrival in Europe of *Homo sapiens,* which we call Cro-Magnons (after the French site where their fossils were first discovered). Neanderthals and Cro-Magnons were bipedal creatures who made tools, used fire, and buried their dead.

Neanderthals appear to have had brains slightly larger than those of today's humans and had thick, shorter leg and arm bones that enabled them to lift perhaps twice as much as the Cro-Magnon people. Theories about their demise include genocide, disease, and lack of efficient insulation and heating to protect them from the numbing cold.

10. *HOMO SAPIENS*

The archaic *Homo sapiens* appeared anywhere from 20,000 to 100,000 years ago in Africa, then migrated into the rest of the Old World. The first fossils of early *H. sapiens* to be identified were found in 1868 in a 28,000-year-old rock shelter near the village of Les Eyzes in southwestern France. Subsequently called Cro-Magnon, the fossils pointed to bipedal beings who used tools and were longer limbed than Neanderthals. Men were generally 5 feet, 6 inches to 6 feet tall and had broad, small faces with pointed chins and high foreheads. Their cranial capacities were up to 1,590 cubic centimeters. Additionally, Neanderthals wielded their tools in a so-called power grip by holding the tool in the palm of the hand with the fingers curled around the body of the tool. By contrast, Cro-Magnons used tools with hafts and shafts. Cro-Magnons had longer life spans than Neanderthals, living well into their fifties, whereas Neanderthals barely lived into their forties.

CHEMISTRY

Prominent Early Chemists

Slowly, alchemy transformed into chemistry as we know it today through the help of colorful and dedicated characters, each with different goals in mind for his work.

1. THALES OF MILETUS

Thales of Miletus (625?–546? B.C.) is the first recorded natural philosopher. Aristotle, the major source for our knowledge of Thales's philosophy and science, identified Thales as the first person to investigate the basic principles—the question of the originating substances of matter—and, therefore, was the founder of the school of natural philosophy.

It is said Thales predicted the solar eclipse of May 28, 585 B.C., and believed that all things originally came from or were composed of water. He died at an old age of heat exhaustion while watching an athletic match.

2. JĀBIR IBN-HAYYĀN

Also known as Gerber, Jābir ibn-Hayyān was a Muslim alchemist, chemist, mystic, and scientist known as the father of chemistry. Born c. 720 and died (possibly murdered) c. 815, he lived 95 years as a member of Isma'ibia, a branch of the famous secret society of Hashhashins, or Assassins, who killed by terrorism and political murders while under the influence of hashish. Jabir's ideas were similar to Aristotle's, as each contained the four qualities of all matter: hot/cold, wet/dry. The complete body of works attributed to Jabir numbers more than 2,000, but researchers have found that grammar, spelling, and references to events are not the same from book to book. Hundreds of years later, after Islam fragmented, authors escaped religious persecution by saying that they had "discovered a lost Jabir book," instead of taking credit for authoring it themselves.

3. AVICENNA

The best known eastern Islamic natural philosopher was Ibn-Sīnā (980–1037), or Avicenna. He wrote more than a hundred books, saying that chemicals maintain their identity in compounds. He was well read, but no one who agreed with him followed his path. One book, *The Cannon of Medicine,* is still used today as a medical encyclopedia.

Avicenna lived at the end of a 300-year golden age of eastern Islamic science. Afterward, the religious and political unity of Islam fragmented. Arabs became suspicious and intolerant. After 1000 A.D., the Turks invaded Mesopotamia and Persia, stopping general learning and cultural exchange. In 1225, the Mongols set back science even further with their invasion, burning texts and destroying learning centers and laboratories.

4. PARACELSUS

Paracelsus (1493–1541) introduced himself as a doctor of medicine; in books, a doctor of theology; in his will, a doctor of liberal arts. Actually, he was a Swiss lecturer, wanderer, apprentice to an assayer, laboratory technician, and studier of medicine. It's assumed that Paracelsus began one of his lectures by burning the works of

Theophrastus von Hohenheim (1493–1541) called himself Paracelsus, meaning greater than Celsus. Although he possessed a very abrasive and egocentric personality, he greatly advanced modern medicine through his denunciations of alchemy and witchcraft.

Avicenna, which would have cost him a large sum in those days. Possessing a paranoid, egocentric, and abrasive personality, he would often use his foul tongue to verbally admonish physicians. He is also reputed to have passed a pan full of excrement around his class and yelled at anyone who left the room. Near the end of his life, Paracelsus cured a magistrate of gout for 100 gulden but received no payment. He tried to sue but lost. He then denounced all officials and spent his life wandering from town to town, offering his services as the physician he had hated. However, Paracelsus greatly advanced modern medicine with his distaste of Galenic recipes and witchcraft.

5. JOHAN BAPTISTA VAN HELMONT

A wealthy Flemish nobleman, Johan Baptista Van Helmont (1577–1644) is credited for the discovery of gases. His troubles with the Spanish Inquisition for his comments on the medical value of saints' bones landed him in prison. He was later kept under house arrest for the last 10 years of his life. While Van Helmont rejected sweating and bleeding as cure, he still used foul Galenic remedies like eye of newt. He was the first to pay attention to gases bubbling from reactions.

6. JOHANNES HARTMANN

Johannes Hartmann (1568–1631), the first regularly appointed professor of chemistry in Europe, served as professor of chymiatria (medical chemistry) at Marburg, Germany, in 1609. He used antimony and mercury preparations but also prescribed Galenic remedies, such as the recipe for a cure for insanity or melancholy: the blood drawn from behind the ear of an ass, holy water, powdered magnet, spirit of human brain, and livers of live green frogs.

7. PIERRE-SIMON DE LAPLACE

A contemporary of Antoine-Laurent Lavoisier, Pierre-Simon de Laplace (1749–1827) was appointed professor of mathematics at École Militaire in Paris at the age of 19. Laplace considered himself the best mathematician in France, and he didn't hesitate to give his opinion on many other subjects, except politics, much to his credit. Laplace had good reason to brag, however, because he and Lavoisier had developed the calorimeter, a way to measure heat, and he contributed to astronomy, giving rise to the analysis of spherical harmonics. Laplace's biggest contribution to mathematics is the differential operator, which describes electrostatics, wave propagation, and heat flow, now called the Laplacian. He also furthered modern gravitational theory and wrote a book on probability theory.

8. JOHN DALTON

John Dalton (1766–1844) began as the son of a poor weaver. He left school at age 11 to work but was so smart that, at age 12, he was hired as a village schoolteacher, teaching boys much older and bigger than he. As a Quaker, he was not eligible for scholarships to British public schools.

Dalton's first discovery was a form of color blindness that he reportedly suffered from himself, now known as daltonism. Originally a meteorologist, he studied mixed gases and the expansion of gases under heat; Dalton's Law is still used to describe the law of partial pressures in chemistry. He lectured on his discoveries in 1803 and published them in *A New System of Chemical Philosophy* in 1808.

Modern atomic theory begins with Dalton's works. He held that all the atoms of an element are exactly the same size and weight and are unlike the atoms of any other element. He also stated that the elements' atoms unite chemically in simple numerical ratios to form compounds.

9. HENRY CAVENDISH

Henry Cavendish (1731–1810) was the first to prove that water is a compound, not an element. He was also a believer in phlogiston, an unseen energy or force that existed in matter, and eventually abandoned chemistry rather than accept the theory that oxygen can be produced from combining acid and metal.

10. MICHAEL FARADAY

As a poor bookkeeper's apprentice, Michael Faraday (1791–1867) grew up reading books that came in for binding and attending labor-class lectures. One day he got a ticket to Sir Humphry Davy's lectures at the Royal Institute. There, he took copious lecture notes, bound them in a book, and shipped it to Davy with a request for a job in chemistry. Davy interviewed and hired him, but to Davy's chagrin, Faraday soon surpassed him. Faraday discovered benzene and coined the terms *ion, anode, cathode, anion, cation,* and *electrode.* In physics, he discovered what is now known as the Faraday Effect, or how magnetics affect light.

Alchemy

Some believe alchemists are quacks—mystics and swindlers and those who claim that they can turn last night's leftovers into gold. Others believe alchemy is not greed but a state of mind and an exploration of the way things tie together. The real alchemy is somewhere between the two, full of scoundrels and scandals, but also of inquisitive thinkers and the foundry of modern chemistry and medicine. And what better way to gain knowledge than in the pursuit of riches?

1. **KNOW THE CODE**

Alchemists often used coded words and phrases to hide their secrets from the general population and competing alchemists. Considered tricks of the trade, one recipe taken from *From Caveman to Chemist* reads "[Place] artificial pearls, mordant or roughened crystal in the urine of a young boy and powdered alum, then dip in quicksilver and woman's milk." Both the boy's urine and woman's milk is in code but gives a connotation more of witch doctoring than a chemical formula.

2. PANTS ON FIRE

Al-Kindi, a Muslim natural philosopher, mathematician, and laboratory technician who lived from 800 to 873, translated and commented on Greek science and philosophy. The man believed that alchemists were charlatans, that silver and gold couldn't be produced from human acts, and even wrote books such as *On the Futility of the Claim of Those Who Pretend the Making of Gold and Silver and Their Deceits* and *Treatise on the Fraudulent Acts of the Alchemists*. But he was also a faker: his *Book on the Chemistry of Perfumes and Distillations* lists adulterants for costly drugs and methods of faking perfumes.

3. GOLDEN HARVEST

The theoretical basis for alchemist transmutation procedures originated in Aristotle's works. The philosopher stated that metals and minerals grew in the ground, like plants, although not as rapidly. Impure metals grew slowly in the ground and turned into gold. Alchemists could obtain large quantities of gold by beginning with impure metal and speeding up the "ripening" process.

4. SCORPIO WOULD HAVE BEEN WORSE

In medieval Europe, early alchemy was related to sympathetic magic, such as voodoo. *Cancer* in Latin meant *crab*. If a man had a tumor, a crab would be tied to the site of the pain or tumor and killed, ridding him of the problem.

5. THE HIGH PRICE OF BEAUTY

Egypt is renowned for its tombs of gold with gold furnishings and utensils for use in the afterlife, and the empire was full of alchemists. Queen Hatshepsut, who lived around 1500 B.C., made a habit of powdering her cheeks with the dust of electrim, a silver and gold alloy. As much as 1.4 million pounds of gold was taken from the desert.

6. **SEALED WITH A KISS**

In the first century A.D. lived Maria the Jewess, or Miriam the Prophetess, who invented the *kerotakis,* an airtight vessel with copper foil suspended at the top, used to heat substances and collect the vapors. The use of such containers in the hermetic arts led to the term *hermetically sealed.* Maria also invented the water bath, or double boiler, to maintain substances at a constant temperature; it is still called a *bain-marie.*

7. **GOOD ADVICE IGNORED**

Writer of alchemy and maker of potable gold, Ko Hung (c. 300 A.D., began his fate as an officer in the military but transferred to study alchemy and medicine with the scholars. Later, he used political connections to be appointed magistrate in a region close to cinnabar deposits that he needed for his chemical studies. On the way to assume the post, Ko Hung passed through some mountains and decided to stop for a while. "A while" turned into the rest of his life; he married Pao Ku, also an alchemist, and stayed there until he died.

In his book, *Pao Phu Tzu,* Ko Hung gives recipes for making gold out of lead, mercury, and other elements, describing their beneficial health properties. Ko Hung also first described the physiological effects of mercury poisoning, a danger later alchemists and chemists frequently faced but never paid adequate attention to.

8. **RECIPE FOR A PHILOSOPHER'S STONE**

Zosimus of Panopolis, a Greco-Roman from Upper Egypt around 300 A.D., described the philosopher's stone and left behind a recipe to make gold. The alchemist began with a substance that had all of the four qualities removed. This prima material would be devoid of all color (black); the alchemist added proper formative qualities in a certain order and proportions to bring the material through the scale of metallic virtues. If the final result was not gold, the alchemist shouldn't despair. The new material could be suffused with "golden essence," holding the power to transform any substance to gold. He described the philosopher's stone as a "stone which is not stone, a precious thing which has no value, a thing of many shapes which has no shape, this unknown thing which is known of all."

9. ALCHEMY FAIRY TALES

Coming out of the Dark Ages, Albertus Magnus (c. 1200–1280), or Albert the Great, roamed the University of Montpellier, in the French Pyrenees, with an unsurpassable reputation in all sciences, including alchemy and magic. A Dominican monk and son of a noble family from Swabia, Germany, Albertus led his star pupil, Thomas Aquinas, through every subject, being surpassed only in botany, by Roger Bacon. While Albertus held a professorship in Paris and Cologne, and a bishopric, he never claimed to be a chemist himself. Albertus was eventually made a saint in 1931.

So great was his reputation that there appeared the legend of Albertus's robot. In his study, the story goes, he kept a bronze statue that Aquinas had cast. One day he gave the figure an elixir, causing it to spring to life. The statue ran errands, swept floors, and did other tedious chores but had one drawback—it talked incessantly. This shortcoming forced Aquinas to smash the robot into scrap with a hammer.

10. THE FIRST BACON REBELLION

Roger Bacon (c. 1220–1292), an English Franciscan friar and Oxford scholar, stressed experimental processes and became an advocate of modern chemistry.

Like many movers and shakers, he was in trouble with the church. Suspected of promoting "dangerous novelties," such as saltpeter (for gunpowder) and advances in optics, Bacon was imprisoned for a time by Pope Urban IV, who placed him under house arrest. In 1284, he was jailed for 10 years. After his death, Bacon acquired the reputation of having been a sorcerer and wonderworker. Because he had speculated about such things as gunpowder, flying machines, telescopes, and mechanically driven carriages, he is celebrated for his foresight.

Unusual Chemists

All living creatures run simple chemical reactions—digesting food, producing wastes, and feeding the muscles (or cilia, or flagella) for movement. Sometimes, however, there's a creature that devises a new way of doing things. Don't envision a squirrel in his tree laboratory mixing this beaker with that. But remember, humans aren't the only chemists on Earth.

1. THE STENCH THAT LAUNCHED A THOUSAND SHIPS

Phoenicians, the businessmen of the Lebanon Mediterranean coast between 1200 and 332 B.C., used a local species of mollusk, *Murex brandaris*, to create purple dye. Each mollusk secreted a minute amount, leaving traders the expensive task of harvesting the shellfish in large quantities when only a small portion of the shellfish was needed. The city Tyre, then the production center for the purple dye, had a perpetual stench surrounding it because of the huge quantities of rotting shellfish. Only the rich could afford purple, and ultimately royal purple became a symbol of nobility.

2. WHO'S THE FATHEAD NOW?

Wounded fathead minnows *(Pimaphales promelas)* discharge a distress signal from their broken skin. Other members of the group flee to safety while predators, such as diving beetles and young Northern pike, move in to feed. But not completely selfless, in their frenzy to feed, predators fight among themselves and churn up the water, creating confusion, allowing the wounded minnows to escape.

3. ONE ORGANISM'S TRASH...

The tobacco plant *(Nicotiana tabacum)* naturally synthesizes nicotine to thwart attack by herbivores and to temporarily store nitrogen. An addictive but trace drug in cigarettes, cigars, and pipe tobacco, nicotine is deadly to humans in large doses.

In 1577, European doctors looking for new cures recommended tobacco for toothaches, falling fingernails, worms, halitosis (bad breath), lockjaw, and cancer. By 1603, English physicians grew upset that tobacco was used by people without a physician's prescription and complained to King James I. The next year, the king increased the import tax on tobacco 4,000%.

4. THE BENEFITS OF BOOKWORMS

In *Theatrum Botanicum,* a 1640 herbal book, John Parkinson noted that books written with ink containing wormwood *(Artemisia absinthium)* were not eaten by mice. He credits this to a second-century Greek philosopher, Galen.

5. EGGS AND DONUTS

When spawning, stick insects attach an extra segment when flicking their eggs to the ground—a capitulum full of fatty lipids. Ants collect this as food and take it back to their nests, where they snip off the capitulum and toss the egg in the ant "dump." There, it hatches safe from predators.

6. OUT OF THE STOMACHS OF BEES

Nectar, a 40% solution of sugar water, amino acids, and other nutrients, encompasses 5% of most plants' biochemical activity. Honeybees *(Apis mellifera)* can gather nearly 50 milligrams, or one-tenth of a teaspoon of sugar per hour. One pound of honey is the nectar from 17,000 foraging trips and 7,000 bee-hours of labor. One hour of a bee's flight burns 10 milligrams, thus making the efforts economical.

Bees collect the nectar in their crop and add digestive enzymes. By the time that the bee returns to the hive, much of the sucrose is converted to glucose and fructose. In the hive, the bee empties its crop into a honeycomb cell or transfers it to another worker.

7. UTILITIES INCLUDED

The cardboard palm *(Zania furfuracea)* in the Cycad family is one of the oldest plants, beautifully adapted with a single species of weevil, the *Rhopalotria mollis*.

For a one- to two-day period, the palm's pollen is mature. During this time, the male cone metabolizes large amounts of starch and lipids at a high rate, amazingly leading to a temperature increase that attracts its pollinators. Weevils bore into the cone; feed on starchy tissue; and lay their eggs to hatch, feed, and pupate. The insects carry pollen to the female cones, which have the same fragrance and starch, but manufacture a poison to keep weevils from eating the seeds.

8. JUST A DROP IN THE BUCKET

The South and Central American bucket orchid *(Coryanthes speciosa)* has devised its own trap in which a specific male pollinating bee falls into a bucket of water and then must try to escape an intricate passageway.

Attracted by the intoxicating fragrance of the orchid, the bee steps onto the flower's slimy surface and falls backward into the bucket formed by the flower's lips, filled with sticky liquid from a gland positioned about the bucket. Unable to fly with its wings stuck, the bee must crawl along a narrow dry passage, brushing against the sides and receiving a pollen sac on its back, which the next flower collects in the same manner. If the bee is a large species, it cannot fit through the passage and will drown.

9. RAFT TO LET

Southern house mosquitoes *(Culex quinquefasciatus)* lay their eggs in large, raftlike masses on water. About 24 hours after being laid, the first eggs emit a chemical compound that serves as a beacon to other females seeking suitable sites. The compound alone would not attract females, but if the eggs and the surrounding water have the odor, the combination works. Thus, mosquitoes find an adequate environment in which to lay.

10. DR. JEKYLL AND MR. HYDE

The ciliate *Lambornella clarki* inhabits the same tree-hole pools as its predators, tree-hole mosquito larvae. The ciliates live as free cells, but when they detect the chemical output of the larvae, the organisms undergo cell division and become parasites. They then enter the larvae, multiply, and kill the hosts, releasing numerous free-form ciliates.

Living Better through Chemistry

Students, don't despair. Chemistry is only as serious as you make it.

1. ANYTHING FOR FASHION

Wool bleachers accomplished their task by spreading wool on a half-sphere wicker frame, then igniting a pot of sulfur underneath, letting the sulfur dioxide fumes bleach the cloth. Worse yet was the linen bleacher's task, which included treading or kneading the cloth in an alkaline solution. Eventually the alkali would eat holes in the skin. Fortunately, cotton was just dipped in sour milk and spread in sunlight, letting the acid peroxide in lactic acid form hydrogen peroxide.

2. DEADLY GOO

"Greek fire" was invented by Callinicus, a Hellenized Syrian and refugee from Maalbek, in the seventh century (673 A.D.). Chemistry with a deadly purpose, petroleum or naphtha was mixed with sulfur and resins to form a jelly that wouldn't evaporate or burn too quickly, and even burn in water. The goo was forced through a siphon and ignited by a flame burning at the tip—the prototype of the flamethrower.

3. ELIXIR OF CHEMISTRY

Before the first alcohol distillations, chemists couldn't dissolve certain substances, such as oil and waxes, in water. After alcohol's discovery, scientists had not only the ability to create more reactions but a source for blue flame extinguishable by water and a great form of intoxication. Chemists worldwide thought so highly of the liquid that they named the distillate alcohol *aqua vitae*, Latin for water of life. Each derivation uses the same meaning: Swedish, *aquavite;* English and Scottish, *whiskey;* French, *eau-de-vie;* and Slavic, *vodka*.

4. HIGH PRICE OF MEDICINE

Bombast von Hohenheim, a.k.a. Paracelsus (1493–1541), a name he called himself, meaning greater than Celsus, had a royal medical following that was against the established Galenist practitioners. Galenists had such a low success rate, as opposed to a high rate of iatrochemists, that nobility began to migrate to Paracelsion physicians, who had been practicing on the poor public. Not willing to risk their lives with the Galenists, the public turned to apothecaries to prescribe drugs. In 1703, apothecaries had the right to practice medicine, a right that was withdrawn in the nineteenth century.

5. PRIORITIES ARE A MUST

German-born Libavius (c. 1540–1616) wrote what is considered the first chemistry book, *Alchymia*, more than 2,000 pages long and including 200 illustrations. The four parts of the book are "Eacheria," or techniques and equipment; "Chymia," or chemical preparations; "Ars Probandi," or chemical analysis; and a section on the theory of transmutation. Also included is a design for a chemical laboratory:

In Europe, apothecaries such as the one depicted here not only prepared and sold medicines but had the right to practice it as well until the nineteenth century.

the ideal "chemical house" contained a storeroom, a prep room, an assistant's room, a crystallization and freezing room, a sand and water bath room, a fuel room, and a wine cellar.

6. **FROGS' LEGS TO LIGHTBULBS**

Luigi Galvani (1737–1798), an Italian anatomist, physician, and physicist, is reported to have inserted a copper hook into a frog's leg and hung it on an iron fence to dry, where it began twitching. Other records say that Galvani was experimenting with the "electrophor" and amber at the same table as his anatomy experiments. Every time a small spark discharged from the conductor to his hand, the frog's leg moved. This led him to discover the electrical influence on muscles. Galvani named the phenomenon "animal electricity."

Alessandro Volta (1745–1827) read Galvani's observation and began his own experiments, finding that he could achieve the same reaction with other animals, saying that it was amusing to make a headless grasshopper sing. Volta exchanged the animal part with paper soaked in brine and amplified the charge with more links. In 1800, he demonstrated his electric battery for Napoleon, who awarded him a gold medal and authorized further experiments in electricity.

7. **A GOOD CATCH**

The youngest of Antoine-Laurent Lavoisier's close associates was Antoine-François Fourcroy (1755–1809), a hardworking but not insightful man whose ancestors had been nobles but whose immediate family was part of the working class. When Fourcroy married, he used part of his wife's dowry to fund a private laboratory where he held private instruction in chemistry.

Fourcroy examined more than 1,000 corpses exhumed from the Cemetery of the Innocents and carefully described the effects of heat, air, water, alcohol, acids, and other solvents on decaying material.

8. AN EXCEPTION TO EVERY RULE

Joseph-Louis Proust (1754–1826) headed a well-equipped chemical laboratory in Madrid, where he decided that each compound has a fixed and invariable composition by weight; that is, each has a formula, which eventually became known as the Law of Definite Proportions. Claude-Loius Berthollet (1748–1822) disagreed, saying that he saw a blurred average of chemical weights in the same compound, but opinion favored Proust. Ironically, some of the compounds that Proust analyzed do have variations and are now called berthollides.

9. WITH A LITTLE HELP FROM HIS FRIENDS

Auguste Laurent (1807–1853), Jean-Baptiste-André Dumas's (1800–1884) assistant, suffered personal attacks because he disrupted the electrostatic bonding theory. The French chemist outworked his contemporaries, discovering several organic compounds. However, he became involved in controversies during France's Napoleonic Era and thus never received a laboratory appointment. So he built one in a basement.

Laurent's nucleus theory was attacked, so he sought the backing of Dumas, who would not support him, claiming that Laurent had misinterpreted his data. Laurent wrote Dumas a sarcastic letter, thanking him for admitting that the credit belonged entirely to him. The two squabbled for the rest of their lives from that point. Dumas later brought out his own theory, the First Type Theory, which barely differed from Laurent's. Still working in organic chemistry, Laurent brought about the Second Type Theory, and finally won over his contemporaries the same year that he died of tuberculosis, leaving a wife and two children destitute.

10. **BEYOND LUCK**

When separating the crystal structures of sodium tartarate and sodium ammonium paratartarate, Louis Pasteur (1822–1895) picked through a group of crystals with tweezers, looking at each one. Luckily, the two crystallize in two different forms at temperatures below 80°F. Had the laboratory temperature been higher, he would have seen two similar crystals and his conclusions would have been in error.

MEDICINE

Cancer Suspects

The answer is, we just don't know all the reasons people are getting cancer these days. Some think it's because humans live longer than they used to. Others say we've complicated life more than before and live in a polluted society compared to our ancestors'.

1. BARBECUED FOODS

Nothing says summertime like food sizzling on the grill. However, that food had better be low fat. Roberta Altman, author of *The Cancer Dictionary*, says that when fatty meats are cooked over heat, the fat drips onto the coals or coils and forms carcinogenic airborne substances, which collect back onto foods through the smoke.

2. HAIR DYE

A study by the American Cancer Society and the U.S. Food and Drug Administration (FDA) published in 1994 found that women who used black hair dyes for more than 20 years had a slightly increased risk of dying from non-Hodgkin's lymphoma or multiple myeloma. They suggested that hair dyes may contain chemicals that alter the structure of DNA, absorbed through the skin and scalp, and darker dyes tend to contain more chemicals. Additionally, a

1993 Harvard School of Public Health study found that women who used hair dyes five or more times per year had twice the risk of developing ovarian cancer.

3. HOMOSEXUALITY

Lesbians have two to three times greater risk of getting cancer, says epidemiologist Suzanne G. Haynes at the National Cancer Institute. Lifestyle factors such as not having children and homophobic attitudes among doctors and nurses, which could be perceived by patients, possibly causing them to stop going for regular checkups, contribute to the increased risk. Both gays and lesbians also may internalize the oppression, which could lead to overeating, smoking, and excessive drinking.

4. MOUTHWASH

You mind your alcohol intake, even consider the alcohol content in cough medicine, but it still may be creeping into your system. A study by the National Cancer Institute found a possible connection between mouthwashes with a minimum of 25% alcohol and oral and pharyngeal cancer.

5. POLYCHLORINATED BIPHENYLS (PCBS)

Used in a variety of industrial applications from 1920 to 1970, including electrical equipment for flame resistance, PCBs have built up in the environment as the result of unwise disposal processes. One of the most hazardous results is in fish, where the chemical can accumulate and harm humans if the contaminated fish is eaten. In 1984, the FDA announced a plan to reduce the tolerance level of PCBs in edible fish from 5 to 2 parts per million, touting liver cancer as a problem. The fishing industry complained, saying that the big cancer-causing questions had not been answered, but by that time, the fishing industry had already been affected.

6. ELECTRONIC USE

Every now and then, we hear another report telling us that technology kills. Cellular phones, alarm clocks, video-display terminals, and more emit high-frequency levels of electromagnetic energy, leading some to believe that this energy could change the structure of the user's DNA for the worse. However, researchers say that all carry too little energy to break chemical bonds or to deposit significant heat in tissue to change DNA. As long as the energy density does not exceed a temperature rise in the skin of 0.1°F, studies cannot prove any effects, and research is marked inconclusive at this time.

7. DDT (DICHLORO-DIPHENYL-TRICHLOROETHANE)

Made most famous by Rachel Carson's book *Silent Spring,* DDT induces tumors and reproductive problems in some animal species, especially eggshell thinning in birds. In 1972, the pesticide was banned in the United States. Later reports followed that DDT wasn't the human hazard described by the media. In 1984, the Word Health Organization and Food and Agriculture Organization of the United Nations issued a report that the chemical, originally thought to be carcinogenic in mice, was not. DDT did induce liver nodules, but those tissues didn't invade adjacent tissues or metastasize. The chemical wasn't inducible in other rodent species, either.

In humans, extra estrogen is seen as a cancer risk. Eggshell thinning, because of its reproductive relationship, was incorrectly linked to "estrogenic effects."

8. NUCLEAR POWER PLANTS

Nuclear industry workers receive an average exposure of 400 millirems of radiation per year. By government standards, 200,000 millirems is found to be hazardous. Residents living 50 miles outside a nuclear plant receive approximately .05 millirems per year. The health risk imposed on those living near fossil fuel plants, compared with those living near nuclear power plants, is well documented.

The result is inconclusive. While radiation levels were high at Three Mile Island and Chernobyl is still the opposite of a vacation spot, scientists say that the people evacuating Three Mile Island were at greater risk from radiation exposure than those who remained in their homes. Three Mile Island never achieved meltdown; the airtight building didn't fail. If meltdown had occurred, the core still wouldn't have melted through the thick concrete base of the core.

9. CITY LIVING

By the numbers, lung cancer occurs more often in cities than rural areas. Some feel that the air is more polluted, that the "natural" life is nonexistent, and that the food is less healthy. Elizabeth Whelm, author of *The Complete Guide to Preventing Cancer*, agrees that the air is different, but lifestyles are different too. She states that cities have more facilities for diagnosis, making the number of diagnosed cases higher.

10. MOST TOXIC SUBSTANCE KNOWN TO MAN

In the mid-1970s, waste oil was sprayed to reduce dust on horse arenas and unpaved roads in the Times Beach, Missouri, area. Horses and other animals in the area died. In 1982, the government evacuated 2,200 residents, spreading panic and fear of the dioxin found in the oil and costing the government $33 million. Soil levels measured 100 to 1,000 times what was believed a safe limit of dioxin ingestion. The substance was earmarked environmental enemy number one.

The panic was traced back to a 1975 rat study in which female rats were fed dioxin excessively. No instances of chloracne, a skin condition caused by the substance, was found in Times Beach, and it was later proved that dioxin binds tightly to the soil, causing no crisis to humans, but being detrimental to foraging and grazing animals.

(The most poisonous element is plutonium metal and all its salts. Its status isn't due to radioactivity, but chemical cell poisoning. A few milligrams will kill a human.)

Cancer Treatments

As of this book's publication, we don't know the recipe for the cure of cancer. However, many healers, running the gamut between doctors and quacks, have suggestions.

1. SHARK CARTILAGE

One of the oldest animals on the planet, the shark has a strong immune system and heals rapidly (a much-needed trait, if you've ever seen them feeding). Sharks produce antibodies for more than bacteria, viruses, and chemicals. Studies show that the shark's all-cartilage makeup combats cancer too. Cartilage doesn't contain blood vessels but makes chemicals that could block angiogenesis, the formation of new blood vessels that tumors develop to feed themselves.

Chinese culture considers shark cartilage an aphrodisiac and elixir of youth, and doctors use parts of the animal when treating burn victims and for inflammatory diseases.

2. COPPER

Your body uses copper for healthy tissue development and function. Studies have shown that in animals certain forms of copper slow tumor growth. However, copper in large doses is not recommended.

3. SOY

We've heard it in the news quite a bit: soybeans and soy products lower risks. In 1963, German scientists isolated a phytochemical in soybeans that blocks the capillaries that bring oxygen and nutrients to tumors, a concept similar to that of shark cartilage. These isoflavones also fight cancer in women by blocking estrogen entry into cells.

4. OZONE

Widely used in Europe for many years, the practice of administering ozone has remarkable effects but little scientific documentation. A highly active form of oxygen, ozone (O^3) gives off monoxide in the body while trying to form a more stable O^2. The monoxide reportedly kills viruses and bacteria while creating an oxygen-rich environment.

5. MAY APPLE

Penobscot Indians in Maine use the rhizome (underground stem) of the flowering plant. Resin was specified in the nineteenth century to treat cancers. Bristol Myers has developed a synthetic derivative called etoposide (VePesid) for treating cancer of the testicles and small-cell cancer of the lungs.

6. ESSIAC

Canadian nurse Rene Caisse used a harmless herbal tea consisting mostly of burdock root, turkey rhubarb root (Indian rhubarb), sheep sorrel, and slippery elm bark to treat thousands successfully from 1920 to 1978, when she died at age 90. Caisse refused payment; she only took voluntary contributions. The formula, Caisse's name spelled backward, missed being legalized in Canadian parliament by three votes in 1938.

7. MISTLETOE

Austrian biology researcher Rudolph Steiner (1861–1925) based mistletoe's healing powers on a hunch. The semiparasitic plant enhances the immune system by stimulating the thymus and regulating cellular immune reactions. Treatment with mistletoe is also called Iscador Therapy. Thought to affect DNA's genetic code, preventing the translation of the gene segments responsible for uncontrolled cell division, the treatment was outlawed by the American Cancer Society.

8. WHEATGRASS

Rich in chlorophyll and nearly identical to human hemoglobin, wheatgrass has 60 times more vitamin C than oranges, and 8 times more iron than spinach. It contains 100 vitamins, minerals, and nutrients as well as all 8 essential amino acids and 11 others. Some feel that the treatment is not primary therapy and only recommend it as a first-phase detoxification.

9. CHAPPARAL

Chapparal, creosote bush or greasewood, is found in the southwestern United States and Mexico. Native Americans brewed the leaves and stems to treat cancer, venereal disease, arthritis, rheumatism, tuberculosis, and colds. Also used by persons with AIDS, chapparal contains nor-dihydroguaiaretic acid (NDGA), which inhibits electron transport in the mitochondria and is also a potent inhibitor of glycolysis.

10. HYPERTHERMIA

Hyperthermia is the application of therapeutic heat to destroy or reduce cancer tumors. The rationale is that cancer cells are more heat sensitive than normal cells. Cancer cells break down when heated over 107°F. The treatment dates back to ancient Egypt. Local hyperthermia involves heating the tumor from 107 to 113°F using ultrasound, microwaves, or radio frequency waves.

During whole-body hyperthermia, a patient is anesthetized and placed in a heated suit or wrapped in thick rubber blankets through which hot water flows. The American Cancer Society removed this treatment from its Unapproved Methods list in 1977. In 1984, it was approved by the U.S. FDA.

Not-So-Common Cold Cures

The "common" cold has plagued humans for thousands of years. In our agony of symptoms, we have devised not-so-common cures, some silly, some uneconomical, and even some unsanitary and life threatening.

1. EXERCISE IN THE NUDE

Oliver Clark, author of *Never Catch a Cold Again,* recommends air baths. These consist of stripping down, doing light exercise and deep breathing, toweling off vigorously, and then reclothing only when the body is warm and comfortable.

2. TRADE A COLD FOR A STROKE

Phenylpropanolamine (PPA), a product found in many cough medicines and diet drugs until November 2000, is believed responsible for 200 to 500 strokes each year from its use. Doctors' first warning sign about PPA came in the 1980s, when medical journals cited several dozen puzzling cases of young women who suddenly had strokes within days of taking appetite suppressants. The FDA issued a warning and suggested a recall, which was followed by a slew of lawsuits. The people at highest risk from using PPA products are woman ages 18 to 49.

3. THE LONG-SUFFERING SHY PERSON

Among other reasons, such as age, income, stress level, and smoking, Jane Brody says that shy people seem to suffer more severe colds than those with more outgoing personalities.

4. HIGH COST OF IMPREGNANCY

Kimberly Clark developed facial tissues impregnated with an antiviral compound to dramatically reduce the spread of colds. Unfortunately for cold sufferers, the tissues were priced three times as high as normal tissues, and the product failed.

5. SPREADING DISEASE FOR SCIENCE

To determine the cause of colds, Dr. Douchez and associates at Columbia University ran a test that successfully passed a "filterable agent" from person to person in a chain of 50 people. Although the experiment was administered in 1924, when less was known about colds and illness, one still must wonder why so many people were needed for the test.

6. GAS ABUSE

Although already considered dangerous in the mid-1930s, chlorine gas clinics allowed cold sufferers to inhale the poison, reportedly to relieve symptoms.

Chlorine gas was first used as a chemical weapon at Ypres, France, in 1915. The immediate effects of the gas include inflammation of the respiratory tract. Irritation of the airway lining leads to a buildup of fluid, which fills the lungs and causes congestion.

7. DRESS FOR THE WEATHER

In 1925, cold prevention was at the top of doctors', mothers', and advertisers' concerns. Lloyd R. Stark quotes the thinking of the time in *The Ultimate Cause and Preventive Cure for the Common Cold,* even mentioning that the height of shoes was cause for concern: "Unequal chilling of the body is a prime factor in producing colds." He cites a draft on the back of the neck, exposure to the wind without a hat, wetting of the feet, and a sudden change from high to low shoes as cold culprits.

8. BAT'S WING, EYE OF NEWT

The Natural Way with Colds and Flu author Penny Davenport gives a recipe for instant cold-free living resembling a witches' brew: "Slice onion and garlic on a flat plate. Cover with runny honey, put a plate on top and leave overnight. The next day, drain the liquid off and take spoonfuls at intervals during the day."

9. A HICKEY TO HEAL YOU

Cupping, a variation on traditional Chinese medicine, uses small jars or cups to stimulate and draw upon the body's energy points. Using 1- to 4-inch-diameter jars, the "doctor" lighted a taper candle in the jar to create a vacuum, then clamped the jar to the sufferer's body at specific points. The jar was left in place for approximately 10 minutes.

10. PLENTY OF R&R

Watching fish or petting a dog or cat can cure your cold, says Carol Turkington in her book *Natural Cures for the Common Cold*. Doctors and mothers worldwide prescribe rest and relaxation to cure the common cold, but Turkington writes that paying attention to your pet "shortens a viral infection, sometimes to an hour." Meditation has also been shown to increase immunoglobulin antibodies.

Fountain of Youth

The mythical fountain of youth was reported to flow with water that cured illness and granted the drinker eternal youth. It was also rumored to sit amid a wealth of gold and silver, which is what historians agree was the magnet that pulled Juan Ponce de León (1460–1521). Unfortunately for him, the fountain couldn't give world peace. Florida natives wounded Ponce de León with a poison arrow during a rebellion, and he died in July of 1521, at the age of 61. The explorer is the first on the list of all those who desperately try to prolong their stay in this world.

1. DON'T DRINK THE WATER, EITHER

In India, Vedic priests used hallucinogenic mushrooms to help them in their tasks. The psychotropic drugs in these mushrooms can pass through the body into the urine. According to Vedic hymns, those who drank the urine of a priest who had eaten sacred mushrooms believed themselves to be immortal.

2. GOES WITH PASTA?

In Chinese practice, sea prawns are believed to build kidney function and increase energy. One recipe says to place two fresh, washed prawns in a wide-mouthed or porcelain jar. Add 250 cubic centimeters of 60-proof spirits and seal the container, and soak for one week. Taken daily, it can be used to cure sexual dysfunction as well as impotence.

3. THE FUNGUS AMONG US

For long life, nothing is better than stewed duck with *Cordyceps sinensi,* also known as Chinese caterpillar fungus. A time-honored tonic in Southern China, the steamed dish is believed to increase vital energy and help eliminate cold sweats, poor limb circulation, impotence, and seminal emissions.

4. TAKE THE BULL BY THE...

Bulls' genitals cooked with *Lycium sinensis,* also known as Jizi, the fruit of the Chinese wolfberry, is said to invigorate the liver and kidneys and strengthen bones.

The recipe calls for the external genital organs, the penis and both testicles, steamed together with the fruit. You may add two pieces of ginger while steaming to rid your kitchen of the rank smell, it notes. Consume both the meat and the juice for treatment of general weakness, kidney problems, back and leg pain, impotence, and excessive nocturnal urination. After all, who could sleep after eating this?

5. TAKE TWO WOMEN AND CALL ME IN THE MORNING

Shunamitism, or rejuvenation of an old man by a young woman, has been the basis of more than a few sitcoms and movies. The practice was actually first seen in the Old Testament, used to vitalize King David (1090–1015 B.C.). A young girl, Abishaq of Shunem, was brought to lie with the king in a nonsexual embrace in his old age. Later the practice rested on the seventeenth- and eighteenth-century belief of the rejuvenating power in the breath or heat of young girls. Scientists postulate that the effects could possibly be a result of undischarged arousal, leading to an increase in sexual hormone production.

6. JAGGED AND ROCKY PATH TO LONGEVITY

In other Oriental medicines, ginseng, ginger, jujube berries, and camphor are combined with the unusual and destructive ingredients of ground tortoise shell (endangering many species); the stomach contents of the musk ox; and "panty," the ground horns of a young spotted deer.

7. LESS IS MORE

Many believe a diet, or lack of diet, increases longevity. Starving yourself, eating very little, or walking away from the table still a little bit hungry may conjure up images of Gandhi or Taoism. One example is Englishman Thomas Parr (1483?–1635), who reportedly lived to the ripe old age of 152. In his one hundred fifty-second year, Parr was invited to visit King Charles I, who encouraged him to eat meat and delicacies. Parr's simple vegetarian lifestyle couldn't cope. Shown by the postmortem, carried out by Dr. William Harvey, the change in diet and the pollution of the city were too much for Parr, leading to his death within a few weeks of arriving in London.

Medical research supports the less-is-more theory in two ways: excess food can encourage various diseases, such as diabetes and heart, liver, and kidney conditions, and minimal diets almost double the life span of experimental animals.

8. FOUR OF FIVE DENTISTS AGREE

Procaine, a nonaddictive synthesized substitute for cocaine commonly found under the trade name Novocain, was reported but never established as an elixir of youth, although support is being revitalized.

German chemist Alfred Einhorn first synthesized procaine in 1905, but it was Dr. Ana Aslan who gave it its famous name. She mixed procaine hydrochloride buffered and stabilized with potassium metabisulfite and benzoic acid, then opened a procaine injection clinic in Romania after World War II, treating famous people, including Charlie Chaplin. The use of procaine declined since the introduction of the local anesthetic lidocaine in the 1940s. The drug, now called GH3, is still prescribed and tested today, mostly in Europe, but is displaced in popularity first by cortisone, then by ginseng. Advocates consider GH3 a provitamin that leads directly to the formation of vitamins in the human body.

9. WHAT'S BAD FOR THE GOOSE IS GOOD FOR THE YOUNG GANDER

Women between ages 12 and 45 need iron supplements during their reproductive years. However, for most men and older postmenopausal women, high stores of iron can lead to heart trouble and cancer by spurring free radicals, reports Jean Caper in *Stop Aging Now!* Careful scrutiny of vitamin and mineral supplements is recommended.

10. THE FOUNTAIN OF FLATULENCE

Chromium, found in brewer's yeast, reportedly helps control insulin and blood sugar. Without this control, excess blood sugar and insulin help to destroy arteries by building plaque in the tubes, which can lead to diabetes, heart disease, and blood glucose intolerance. Recommended use of brewer's yeast, originally a by-product of beer brewing, is four tablespoons per day, but the substance has been associated with excess gas.

TECHNOLOGY

Inventions from Outer Space

Since 1958, NASA has sent the first humans to the Moon, explored eight of nine planets, and set up the first reusable space transportation system. The technology that makes space programs possible has also inspired the development of inventions for Earth-bound mortals. Space technology has been applied to health, medicine, recreation, construction, and manufacturing.

1. NUCLEAR MAGNETIC RESONANCE SCANNING

Toward the end of the 1960s, a NASA satellite called *Landsat* sent images of Earth to receiving stations on the ground. On each successive orbit, the satellite built up a sequence of image strips that covered the globe. Detectors recognized unique "signatures" of various features—crops, water, buildings—in a system called thematic mapping. Dr. Michael Vannier, who began his career with NASA, pioneered the use of nuclear magnetic resonance (NMR) scanning to determine physical ailments. Unlike X rays, NMR does not expose the body to radiation.

2. ANTIFOG SPRAY

In 1966, aboard the *Gemini 9,* Eugene Cernan expended more energy than he had planned to as he tried to reach a maneuvering unit in order to rendezvous with another target. Because he was sweating profusely, his faceplate fogged and eventually caused him to abandon the operation. After the flight, NASA developed an antifogging spray from liquid detergents, deionized water, and a fire-resistant oil. In 1980, Tracer Chemical marketed it for industrial and consumer use. Customers include fire departments and car-window manufacturers.

3. SELF-HEALING COMPUTERS

Lengthy space missions to outer planets could take decades, so NASA developed unmanned probes. However, radio control didn't work fast enough, so robots had to make independent decisions. From a special research program called Self-Test and Repairs (STAR), self-repair systems arose. Intelligent machines interact with humans to monitor routine work functions. Aircraft builder Lockheed Martin incorporated STAR technology into the super-stealth F22 fighter called *Raptor,* which can reassemble its components if knocked out.

4. KEYBOARDS FOR DISABLED WORKERS

Astronauts and pilots control complex machines while making split-second decisions. NASA developed touch-screen loading and single-key input boards to simplify complex tasks. Computerized information displays replaced individual instruments. In cooperation with Infogrip Inc. of Baton Rouge, Louisiana, the Stennis Space Center and Mississippi State University developed the BAT chord keyboard with just seven keys and a simplified finger system. The keyboard can be operated single-handedly, enabling disabled users full participation in computerized tasks. It is also suited to those with impaired vision because it is based on braille.

5. RADIATION BLOCKERS

Sunglasses with enhanced radiation-blocking effects evolved from NASA's research in the 1970s on advanced optical coatings for mirrors and lenses for cameras and lenses used in space. NASA employees Laurie Johnson, Paul Diffendaffer, and Charles Youngberg produced a filter for protecting eyes with a dye curtain that filters out harmful blue and ultraviolet light.

6. SNEAKER GEL

Aerospace engineer Al Gross of Lunar Tech Inc. in Aspen, Colorado, was asked to design an advanced athletic shoe that offered increased shock absorption over an extended lifetime. He decided to investigate how NASA used bellows in the joint areas of pressurized spacesuits. His research led him to produce an external pressurized shell with horizontal bellows that replace foam materials as cushions.

7. INDESCTRUCTIBLE IDENTIFICATION

To inventory space shuttle components, NASA and CiMatrix Corporation developed in 1997 a microetching system that didn't require too much surface area to install a label, such as in bar-code systems, and could withstand tremendous heat. The tag system's indestructible identification has been laser etched on steel, plastic, paper, fabric, ceramics, and composites. Items marked by the system are inscribed for life, and their identity can never be erased.

8. WEATHER-RESISTANT COATING

When the *Apollo* rockets were ignited in the late 1960s and early 1970s, the Cape Canaveral launch center needed protection from thermal shock so that it could be reused. NASA scientists pushed forward ceramic chemistry to produce a coating that would withstand not only tremendous heat but also the salty effects of the nearby Atlantic Ocean. The result was IC 531, a quick-drying ceramic coating.

Since 1982, Inorganic Coatings, Inc., has produced the coating, which has been used on the Golden Gate Bridge, the Po Lin Buddha, and the Statue of Liberty.

9. MAGNETIC LIQUIDS

NASA felt that it couldn't rely on magnetic fluids to provide adequate fuel injection in a weightless environment, so it developed thrusters to do the job. Two NASA scientists, Dr. Ronald Moshowitz and Dr. Ronald Rosenweig, created a leakproof seal for the rotary shaft of a system used in making semiconductor chips. Most computer disk drives now employ magnetic fluid exclusion seals. The fluids are used in motorized shafts designed to operate in a vacuum chamber or an ultraclean environment.

10. ELECTRONIC WATER FILTER

During the *Apollo* flights, astronauts drank water generated as a by-product of the craft's electrical energy system. Hydrogen and oxygen produced electricity in fuel cells and combined to make water used for cooling and drinking. Water from the fuel cell was purified by an electrolytic water filter. Stemming from environmental concerns of the 1970s, Paul Pedersen, president of Western Water International, produced, in conjunction with NASA's technology, an ion-based water-purification system that eliminated lead in water systems.

Artificial Intelligence

Quite a few researchers refuse to admit limitations of current technology. Here are a few projects aimed at building intelligent machines, capable of communicating in natural language and acquiring new knowledge.

1. THE TURK CHESS AUTOMATON

From 1770 to 1854, a series of chess players faced a fierce-looking turbaned puppet seated at a cabinet with a chessboard. Behind the box, the player was shown an array of gears and springs. Chess players were impressed by the Turk, as the puppet was called, and the puppet usually won. However, jammed into the closet with the puppet was a human chess player who observed opponents by candlelight and operated the Turk's responses.

2. BABY HAL

Despite many efforts to build a facsimile of HAL 9000 of *2001 Space Odyssey* fame, talking computers, or "chatterbots," remain immature. However, Artificial Intelligence NV of Boston and Tel Aviv created its own HAL, a truly conversant computer. HAL successfully passed an adapted Turing Test for 15-month-olds, a test that assumes that a machine is intelligent if it can fool human observers into thinking that they are communicating with another human being. Reviewing transcripts of HAL's conversations, a child-development specialist declared him a healthy, normal little boy. HAL reportedly talks like an 18-month-old now. Contrary to the traditional set of hardwired rules and a vocabulary database, AI developed HAL from a behaviorist approach—language is a skill learned through imitation and rewards. Lead scientist Jason Hutchens predicts that HAL should reach adult speak in approximately 10 years.

3. WEBMIND AI ENGINE

Webmind, Inc. (a.k.a. Intelligenesis) developed Webmind AI Engine, intended to display true human-level intelligence. Unfortunately, the company announced bankruptcy in 2001. It had planned to create a program that can hold intelligent (though not necessarily human-like) English conversations. The idea behind the program was the "psynet model," which held that the Internet of the future will be an immensely intelligent system with powerful synergies with human intelligence. Software releases would culminate in a conversational system.

4. STARLAB

Another blue-sky research laboratory, this time in Brussels, Starlab was the home of Professor Hugo de Garis and his Starbrain. The original aim of the project was to build an artificial brain with a billion artificial neurons by 2001, using "evolved" cellular automata-based neural circuit modules. This project also went bankrupt.

5. COMMONSENSE

The Open-Mind Commonsense Web site encourages users to answer simple questions, describe the relation between two words, describe images—all in order to build a database of commonsense facts. The site encourages visitors to teach a computer to recognize speech or handwriting.

6. BRAINHAT

Brainhat Corporation of East Hartford, Connecticut, developed a natural-language operating system that can be programmed in English, serve as a clearinghouse for English-based events, interact with users, and direct the operation of robots. Developer Brian Dowd says that the operating system addresses the need for knowledge-based "conscious computing" that can replicate human thought more closely than traditional sequential programming.

7. AIBO

One of the first entertainment robots, Aibo was considered the pet of the future. Sony's toy can walk, play, sit, and stretch. The company claims that it has emotions, instincts, and the ability to learn. The cyberpet shows emotions through sounds, motions, and eye lights. A happy Aibo might sing tunes or even dance. Aibo also has four instincts: love, search, movement, and recharge. When slapped on the head frequently, Aibo will learn not to repeat an action, but it will learn to repeat behavior that is rewarded with gentle touches to the head. The robotic dog sold out within 20 minutes after being offered.

8. ROBONAUT

NASA's space-walking automaton, or android, could ideally construct and repair space stations. A 1990 study concluded that an orbital station would require 75% more space-walking time than originally planned. NASA thus looked to its Dextrous Anthropomorphic Robotic Testbot, or DART, which had two arms and two hands. But DART was too bulky. So the agency is building Robonaut, designed to be the size of a suited astronaut with the same dexterity. In 2000, the NASA team installed Robonaut's left hand and torso. It's predicted that it will be ready for launch in 2003, complete with a sense of touch.

9. MINDPIXEL DIGITAL MIND

The Mindpixel Digital Mind Modeling Project claims to be the world's first program attempting to become artificially conscious by talking to Internet users. Its goal is to collect 1 billion Mind Pixels—some true or false statements—over 10 years.

10. STARBRAIN PROJECT

The aim of the Starbrain Project is to build an artificial brain that will control the behavior of a life-size robotic kitten. The brain will contain approximately 75 million artificial neurons that live in a specially built computer. The kitten will see, hear, and feel. Its behavior will be remotely controlled by the artificial brain. Behaviors will evolve, rather than be programmed. The brain of the robotic kitten is the CAM Brain Machine (CBM), which consists of 72 microprocessors, named field programmable gate arrays, which don't have fixed logic circuits. Neural networks are grown and evolved using genetic algorithms. Conducted at Starlab, a privately funded research company based in Belgium, the project's leader, Hugo de Garis, plans to complete the kitten by 2002.

Your Flight Has Been Canceled

The greatest manned space flights that never were is a sad saga of perished crews, crashed space stations, and canceled programs.

1. SUBORBITAL FLIGHTS

The original *Mercury* project envisioned all the astronauts making an initial suborbital hop across a Redstone booster before making an orbital flight aboard an Atlas booster. However, the plan was delayed by Yuri Gagarin's orbit. The sinking of Gus Grissom's capsule and Gherman Titov's full-day orbital flight in August 1961 made U.S. suborbital flights look pathetic. Further suborbital flights were canceled, and John Glenn rode into history.

2. SOYUZ

The first Soyuz flights were designed to prepare for manned rendezvous, docking, and crew transfer, which were accomplished on *Soyuz 4* and *Soyuz 5*. Cosmonaut Vladimir Mikhailovich Komarov piloted the *Soyuz 1*, which was first launched in August 1961, with the second flight to depart the following day. The solar panels of Komarov's spacecraft failed to deploy soon after launch, and the second team was to rendezvous with *Soyuz 1* to fix the panel. However, the launch of *Soyuz 2* was canceled due to heavy rain. Low on power and battery reserves, Komarov tried to land the following day but crashed when a parachute failed to deploy, killing him. After the disaster, it was discovered that *Soyuz 2* had the same problem. If it had launched, it probably would have docked successfully but would have crashed on landing, killing four cosmonauts instead of one.

3. *MERCURY* TEST

After booster problems on the *Mercury MR-2* test flight, Werner von Braun insisted on another unmanned test. So a Mercury capsule was launched on a flawless test on March 24, 1963. If NASA had overruled von Braun, the manned *Freedom 7* capsule would have catapulted Alan Shepard into the history books as the first man in space, beating Yuri Gagarin's flight by three weeks.

4. *GEMINI 9*

Elliot See and Charlie Bassett were scheduled to be the prime crew for *Gemini 9*. On February 28, 1966, they were flying in a NASA T-38 trainer to visit the McDonnell plant in St. Louis, where the spacecraft was in assembly. See misjudged his landing approach, pulled up from the runway, and hit the building where his spacecraft was being assembled. Both astronauts were killed. The resulting crew reassignments determined who would be the first man on the Moon.

5. **SKYLAB 3**

Influenced by the stranded Skylab crew portrayed in the movie *Marooned*, NASA in 1973 provided a crew rescue for the only time in its history. If trouble developed, a rescue CSM would be launched to rendezvous with the station. During Skylab 3, a thruster of the

Apollo service module developed leaks, and astronauts Vince Brand and Don Lina began preparing for a rescue of astronauts on board the station. However, the problem was solved on board, and the first space rescue wasn't necessary. The crew returned safely after the end of its 59-day mission.

6. **EXCESS HARDWARE**

After the Apollo, Skylab, and ASTP programs were completed in 1973, NASA realized that it had surplus hardware on its hands: two Saturn V and three Saturn 1B boosters, one Skylab space station, three Apollo CSMs, and two lunar modules. NASA therefore considered an international Skylab, using a Saturn VSA-514 to launch a second workshop module and international payloads. The station would be served by Apollo and Soyuz, then by the space shuttle. However, funds weren't forthcoming, so the hardware was mothballed in August 1973. In December 1976, the booster and spacecraft were given to museums. The opportunity to launch an international space station at lower costs and 20 years earlier was lost.

7. **PRESHUTTLE**

In late 1979, shuttle orbital missions were due to start, and the orbiting Skylab space station wasn't expected to decay until 1983. In a plan to save Skylab, a small reboost module would dock with Skylab to boost the station to a higher orbit. However, the shuttle program was hit with delays; before the first shuttle flew, Skylab burned up in the atmosphere and crashed into the Australian outback on July 11, 1979.

8. **FEMALE COSMONAUTS**

In September 1986, the Russians planned to send an all-female crew to dock with Mir. The flight, to be launched on International

Woman's Day, was canceled because of the birth of Cosmonaut Savitskaya's baby. No female cosmonauts would be trained again until a decade later.

9. *CHALLENGER*

After the *Challenger* disaster, shuttle missions were canceled in 1986 and 1987. The missions included deployment of the Ulysses spacecraft, commercial communications satellites, and the Hubble Space Telescope.

10. **BURAN**

Buran-Energia was Russia's answer to the American space shuttle program. Envisioned before the American shuttle, yet built after *Columbia* blasted off, the Russian project was one of the largest, most ambitious, and most expensive space programs attempted. Yet Buran-Energia had only one successful unmanned space flight before the project was suspended for lack of funds in 1993. Because of the powerful Energia core booster, the Buran shuttle did not need an engine. There was talk of reviving the program in 2001 for space tourism and a possible trip to Mars.

Computer Advancers

You could say that Alan Turing invented the computer in 1937, or that Douglas C. Engelbart invented the mouse in 1963, but to use a dreaded tobacco slogan from Virginia Slims, You've come a long way, baby! Here are the top 10 things that have made computers into the necessary items they are today.

1. MAINFRAME

A large computer system that handles a handful to several thousand online terminals, large-scale mainframes can have gigabytes of memory and terabytes of disk storage. Users connect to communications networks through front-end, "dummy" processors. Mainframes allow shared access to immense amounts of data within a corporation.

2. THE PC

Personal computers (PCs) defined the machines for home use, following a levelist theory of the computer world—give affordable machines to everyone. The PC evolved from a "techie" toy to a mainstay in the lives of millions through constant innovation over the last 20 years.

The first IBM PC was introduced in 1980. There were many other systems before, such as the Radio Shack TRS-80, Apple, and CBM PET 2001, which transferred computing power from big mainframe machines to small stand-alone systems. The development of the PC led to the demise of many mainframe computers and a whole generation of programmers and engineers who grew up with them.

3. GRAPHICAL USER INTERFACE

A graphics-based user interface incorporates icons, pull-down menus, and a cursor to provide an intuitive design to allow users to run applications. Although it is not pioneered by Macintosh or Windows, both systems are examples. Prior to the interface's development, programs were run on a computer via a prompt such as DOS.

4. MOUSE

The mouse or trackball is used to access functions or to help users draw an image. It allows easier interaction between computer and human and reduces constant use of the keyboard.

5. HARD DRIVE

A hard drive is a permanent, nonvolatile memory that holds its contents without power. Its development stopped the user from having to start from scratch with each use of his or her computer. Over time, hard drives have grown exponentially in size, practically eliminating the need for removable disks beyond transporting data.

6. WORD PROCESSING

The creation of text documents on computer killed the typewriter because of the ease with which documents can be edited, fonts interchanged, and documents reprinted.

7. E-MAIL

The transmission of messages between computers, electronic mail requires a messaging system, which provides a store and a forward capability, and a mail program that provides the user interface with its send and receive functions.

In addition, the advent of attachments matured the technology from just exchanging text to exchanging graphics and showed people that e-mail was more than just a pastime.

8. THE INTERNET

Originally adopted by the military, the Internet is most widely used for academic and commercial purposes. In 1994, an estimated 30,000 interconnected networks worldwide were on the Net. This number was expected to double each year.

Like e-mail, the Internet gave a purpose beyond word processing to the average person, who gained an exchanging place for knowledge. Advertising increased the hype of the Internet, making skeptics pay attention, with the hint of becoming rich.

9. MULTIMEDIA

Multimedia is defined as anything that disseminates information in more than the plain text form. Multimedia includes the use of audio, graphics, animation, full-motion video, and video conferencing. It definitely made computers more "fun."

10. HIGH-CAPACITY REMOVABLE STORAGE

High-capacity external storage, such as disk or tape, allowed users to save massive amounts of data to a transportable medium, such as a Zip or SyQuest disk. This made computers more useful to businesses and home users.

Internet Accelerators

Born an ugly text baby used for simple information exchange, the Internet has matured into a scholar, news anchor, street hustler, photographer, and innumerable other professions. But it didn't make it on its own.

1. **HTML**

Hypertext markup language, a standard for defining hypertext links between documents, was built primarily to link content but now serves as a foundation for browsers to access and display pages.

2. **E-MAIL**

Who doesn't like to get mail? (Not the junk kind, of course.) E-mail has surpassed postal mail, costs less than long-distance telephone calls, and provides instant messaging for those needing a reason to "get wired."

3. **DOMAIN NAMES**

Like telephone numbers, domain names are the unique names that identify an Internet site. Originally a set of four sets of numbers, the use of actual font changed Web sites to identifiable names, increasing many sites' commercial value.

4. BROWSERS

Graphical user interfaces assist in navigation and hold the hands of those unwilling to jump. Browsers such as Internet Explorer, Netscape, and Opera make content more accessible from the desktop, as compared to previous telnet browsers such as LYNX.

5. ISP

Internet service providers handed Internet access to the public.

6. CGI

Common gateway interface is a general scripting that allows developers to use logic when building Web pages, using the power of a computer to create and display content for the Web.

7. SSL

Secure socket layers provide encryption between the client's browser and a server. This opened the door for secure commerce and the transmission of sensitive data on the Internet.

8. WYSIWYG EDITORS

What-you-see-is-what-you-get editors allowed the development of rapid application. While not widely accepted by serious developers, WYSIWYG allows the novice user to publish content on the Internet and sustained the desktop publishing revolution in the 1990s. The old definition refers to text and graphics appearing on the screen the same way they print.

9. SEARCH ENGINES AND PORTALS

Both influence, some say cram, people onto the Internet. Portals such as Yahoo!, AOL, and MSN strive to be your home base for interactions on the Internet.

10. **WIRELESS INTERNET**

The Internet enabled wireless devices such as cellular phones, personal digital assistants, and household appliances to be at the forefront of communication technology. Using the wireless Internet, one day your refrigerator will call in its own repair.

Most Fascinating Systems

The most fascinating systems are defined not as the most interesting, but the most in-depth, extreme systems that if they went down, daily life would cease. Often these systems have several parts, each developing subsystems of its own, making a failure impossible.

1. STOCK EXCHANGE

When you engage a broker, he or she relays your trade to the floor of the appropriate exchange, and a representative of the company (or, more commonly, a computer representing the company) makes the trade on your behalf. Computers may execute your order with little or no human intervention. You can log on to a brokerage firm's Web site, enter an order, have the trade executed, and receive a confirmation within 60 seconds or less.

Because large numbers of outlets are available to traders, a crash in the system is not likely. If one computer crashes, the user just gets on another machine. The various stock markets use a self-regulatory system in which they investigate and adjust all customer complaints, standardize practices between firms, and make sure that all the member firms observe all federal and state laws.

2. POWER GRID

The power grid is a vast system of high-voltage lines crisscrossing the countryside, designed, built, and operated over the years by electric utilities to serve homes and factories within each utility's geographic reach.

In a *Los Angeles Times* article, Charles Piller suggests that technology criminals can jump from Internet-connected networks to grid-control systems. Deregulation of the energy industry has led to the development of numerous online energy-trading systems, which experts say are less secure than computer networks maintained by utility companies. If a power-control network is breached, a hacker can black out entire cities, and even cause physical damage at energy plants. In spring 2000, the California Independent Service Operator computer networks, which regulate much of the electricity flow in the state, were hacked into during the energy crisis.

3. AIR TRAFFIC CONTROL

During peak air travel times in the United States, there are about 5,000 airplanes in the sky every hour. The United States divides its airspace into 21 zones (centers), with each zone further divided into sectors. Also within each zone are portions of airspace, about 50 miles in diameter, called terminal radar approach control (TRACON) airspaces. Within each TRACON airspace are a number of airports, each of which has its own airspace with a 5-mile radius.

The air traffic control system, which is run by the Federal Aviation Administration (FAA), has been designed around these 21 divisions. It is further divided into sections, the top being the Air Traffic Control System Command Center (ATCSCC), which oversees all air traffic control. It also manages air traffic control within centers where there are problems (bad weather, traffic overloads, inoperative runways).

As an aircraft travels through a given airspace division, it is monitored by one or more air traffic controllers responsible for that division. The controllers monitor the plane and give instructions to the pilot. As the plane leaves that airspace division and enters another, the air traffic controller passes it off to the controllers responsible for the new airspace division.

4. ARPANET

The Advanced Research Projects Agency Network, or ARPANET, was the network that became the basis for the Internet. It was funded mainly by U.S. military sources and consisted of a number of individual computers connected by leased lines.

In the 1980s, ARPANET was replaced gradually by a separate new military network, the Defense Data Network, and NSFNet, a network of scientific and academic computers funded by the National Science Foundation. In 1995, NSFNet began a phased withdrawal to turn the backbone of the Internet over to a consortium of commercial providers, such as PSINet, UUNET, ANS/AOL, Sprint, MCI, and AGIS-Net99.

5. DOMAIN NAME SERVERS

The domain name server (DNS) system forms one of the largest and most active distributed databases on the planet; without it the Internet would blink to a halt.

DNSs translate the human-readable domain name into the machine-readable Internet provider address. During one usage, users may access the domain name servers hundreds of times. No other database on the planet gets this many requests.

Because all of the names in a given domain need to be unique, a single entity controls each list and ensures no duplicates. For example, Network Solutions maintains the .com list. When you register a domain name, it goes through one of several dozen registrars who work with Network Solutions to add names to the list. In turn, Network Solutions keeps a central database known as the "whois" database that contains information about the owner and name servers for each domain.

The DNS system is a distributed database. Every domain has a domain name server somewhere that handles its requests, and behind each a person maintaining the records. The list is completely distributed throughout the world on millions of machines administered by millions of people, but looks like a single, integrated database.

6. CREDIT CARD VERIFICATION NET/CREDIT REPORTING COMPANIES

Made apparent in the movie *Fight Club,* destroying all history of credit would be devastating, yet liberating to a great many. Credit Card Verification System (CCVS) is software that lets merchants process payments. Under the credit system, the bank credits the account of the merchant as it receives sales slips and then assembles charges to bill to the credit card holder at the end of the billing period.

The first credit card was BankAmericard, which was started on a statewide basis in 1959 by the Bank of America in California. Beginning in 1966, the system spread to other states and was renamed Visa in 1976.

In order to offer expanded services, such as meals and lodging, many smaller banks that earlier offered credit cards on a local or regional basis formed relationships with large national or international banks.

7. ATOMIC CLOCK

An atomic clock uses the resonance frequencies of atoms as its timekeeper. This means that they are extremely consistent. In 1945,

Columbia University physics professor Isidor Rabi envisioned the concept, and by 1949, we had the first atomic clock.

In the United States, the standard of time is regulated by the U.S. Naval Observatory's Master Clock (USNO), the official source of time for the Department of Defense. Without atomic clocks, global positioning satellite navigation would be impossible, the Internet would not synchronize, and the position of the planets would not be known with enough accuracy for space flight.

8. SOCIAL SECURITY LISTING

The original and essential purpose of Social Security numbers is to keep track of the money flowing into the Social Security program. When someone dies, his or her number is simply removed from the active files and is not reused. In theory, a time would come when "recycling" numbers may be needed, but not for a long while. The nine-digit number allows about 1 billion possible combinations, and to date a few more than 400 million numbers have been issued.

The first three digits are assigned by the geographical region in which the person was residing at the time a number was obtained. Generally, numbers were assigned beginning in the northeast and moving westward; that is, people on the East Coast have the lowest numbers. The remaining six digits in the number are more or less randomly assigned and were organized to facilitate the early manual bookkeeping operations associated with the creation of Social Security in the 1930s.

9. FEDERAL RESERVE

The Fed, with 12 regional federal reserve banks, uses three monetary policy tools to influence the availability and cost of money and credit: open market operations, the discount rate, and reserve requirements.

The Federal Open Market Committee (FOMC) meets eight times a year in Washington, D.C. For each session, economists at the board of governors and the reserve banks prepare extensive economic analysis of statistics from every region and industry in the country. The FOMC also gets grassroots information from the boards of directors of the reserve banks. For each meeting, reserve bank presidents bring reports on their districts' economies, based on information from reserve bank directors, other district residents, and the banks' research departments. These reports are part of the briefing materials used by the FOMC in formulating monetary policy.

Twice a year, the chairman of the board of governors reports to Congress on the FOMC's economic views and projections, as well as the issues likely to affect near-term monetary policy decisions. The long-term goals of the Fed remain constant: a strong economy with stable prices, maximum sustainable employment, and opportunity for economic growth.

10. **AIRLINE RESERVATIONS**

In the United States, four giant airline computer systems, known as CRSs, or computer reservations systems, handle nearly all the airline reservations in the country. Although each airline has a "home" CRS, the systems are interlinked so you can buy tickets for any airline from any CRS. The dominant systems in the United States are Sabre (American and US Airways), Apollo/Galileo (United), Worldspan (Delta and Northwest), and Amadeus (Continental and many European lines). Independent airlines usually have their own sites to buy tickets. In theory, all the systems show the same data. But in practice, they get a little disharmonious.

Hacked Off

Crackers, or hackers, as they hate to be called, have hacked into enough Web sites to get them imprisoned for life. Those who do pay the penalties, such as the most famous computer bandit, Kevin Mitnick, often have an outside following hacking for the spirit of those incarcerated. Mitnick's very public story is number one on the top 10 list of famous computer thievery, but here are the 10 runners-up, derived from Matt Lake's *The 10 Most Subversive Hacks of All Time*.

1. WORM ATTACKS

In November 1998, Robert Tappan Morris, the 22-year-old son of a security expert for the National Security Agency (NSA) and a Cornell University graduate student, wrote a benign program to map every server on the Internet.

Known as The Worm, the program was supposed to visit a server, copy itself onto it, and move to the next. However, a misplaced decimal point in the code made The Worm copy itself not once but indefinitely on each server, crashing more than 6,000 servers and stopping the Internet for that day.

Morris was sentenced to three years of probation, community service, and a $10,000 fine. Conspiracy theorists thought that Morris was covering for the NSA, since the incident showed how defenseless the Internet was.

2. GO HACK YOURSELF

The 1995 movie *Hackers* received less-than-rave reviews from the cybercommunity—so much that the movie's site was hacked into to show a scathing parody of a movie review. MGM/UA kept the hacked site live, still found at www.mgmua.com/hackers/inventory/hacked. Some believed that the stunt's purpose was to promote the movie.

3. DUCK WORLD

Internet publicity was also considered when *Jurassic Park: The Lost World*'s site (www.lost-world.com) was hacked into shortly before the movie's release. The hack replaced the movie's trademark dinosaur with a duck and the words "Duck World, Jurassic Pond."

Within a day, online magazine *Beta* discovered that the duck image was a professional image map with a time stamp two days earlier than the original *Lost World* graphic.

4. U.S. VS. INDECENCY

In 1996, the U.S. government attempted to control Internet content considered harmful to minors, specifically pornography, through the Computer Decency Act (CDA). Masses cried censorship, but only a few messed with the Department of Justice's Web site to protest the law.

Early on Saturday, August 17, 1996, system administrators of the Department of Justice site discovered that the site had been hacked into and began two-day repair operations. Meanwhile, visitors were treated to parodies of the Justice Department's statements about the CDA, which was eventually overturned by the Supreme Court.

5. SEEING RED

The Ghost Shirt Society attacked Kriegsman Furs and Outerwear in November 1996, changing the site's front page from an image of a white fox fur coat with the company's slogan to a monochrome picture of a similar fur coat daubed in red and the words "fur is dead."

Unusually, the perpetrators didn't brag about how they "owned" the site's administrators and even somewhat apologized to the technical staff, saying that the hack was done with animals' well-being in mind. On the hacked page, they left links to animal-rights sites.

No one was ever caught; the hack is remembered as leading the way for hacking as a form of nonviolent protest.

6. ATTACKING THE PUBLICIZERS

The defacer known as Fluffy Bunny defaced www.attrition.org, a site that points out such attacks. Replacing the main page with a single image of the pink bunny, the person (or persons) is also known for defacing the SANS Institute (www.sans.org), Exodus Communications Security Page (www.security.exodus.net), Apache, SourceForge, Stileproject, and possibly others.

Attrition.org posted its own hack with the text: "The Fluffy Bunny demands that attrition mirror all of his defacements or your network will be overrun with more bunnies than Australia!" found at http://attrition.org/mirror/attrition/2001/07/28/www.attrition.org/.

7. OPERATION DNS STORM

When Network Solutions (InterNIC) began charging $100 to register domain names in 1995, many computerites were hacked off. One of the few alternatives to the InterNIC was AlterNIC, the brainchild of Eugene Kashpureff. AlterNIC offered a different way to register domains and used alternatives such as .ltd and .sex.

In July 1997, Kashpureff diverted traffic from Network Solutions to AlterNIC, dubbing the feat Operation DNS Storm. The illegal move forced Kashpureff to flee to Canada, but he was eventually arrested and found guilty of computer fraud. His partner is still providing an alternative.

8. SICK JOKES ABOUND

After several planes crashed, ValuJet Airlines renamed itself AirTran in 1997. Its new Web site featured a banner headline, "The Making of a New Airline," and a press release announced the changes.

Hackers couldn't resist the site and quickly filled the pages with sick humor. The banner headline was replaced with: "So we killed a few people. Big deal." AirTran promptly removed the hacked page. Although the hackers were never caught, they sent a copy of the page to *2600 Magazine* for posterity.

9. MORE POSTERITY

A group calling themselves HFG (Hacking for Girliez) replaced the main page of the *New York Times*'s site on September 13, 1998, with attacks against writers working on a book about hackers. To the average person, this hack looked like gibberish. Those who deciphered the hack read that HFG flaunted its own abilities, as well as raised consciousness in support of famous hacker/cracker Kevin Mitnick; *Times* writer John Markoff helped bring Mitnick to national attention and also coauthored a book with Mitnick's capturer, Tsutomu Shimomura. The *New York Times* fixed its site, but the perpetrators were never caught.

10. ATTACKING THE ATTACKER

The United Loan Gunmen attacked Matt Drudge's Drudge Report on September 13, 1999. Except for a change to the site's banner, which the hackers changed to the ULG Report, the front page maintained its general appearance.

The headlines were changed to mention issues such as "Kevin Mitnick still in jail," but the site still contained Drudge's regular column and archives. The site was fixed an hour after the hack. Although the hackers were never found, the group claimed hacks on ABC and C-SPAN.

Dot.bombs

Newsfactor.com reported on several of the biggest losses in the Internet downturn in 2000–2001. The biggest losers were the venture capitalists, which had their coffers open in many different sites, often with two or three failing at a time.

1. **CMGI**

Venture capitalist and net incubator CMGI's losses were a reported $2.562 billion in the second quarter of 2000. CMGI expected to nurture more than 70 dot-coms.

The company's founder, David Wetherell, said that bad dot-com wagers were made by CMGI's @Ventures unit, including Furniture.com, which closed in November 2000, and has since filed for bankruptcy after burning through more than $45 million. Mother Nature.com, another @Ventures company, closed in November after posting a $6.7 million loss in its last quarter. BizBuyer.com, which closed in December after running through $70 million, was also a CMGI investment. While the company owned Altavista.com, it sold subsidiary Raging Bull, an Internet stock discussion site.

2. PETS.COM

This dot.bomb lost 19¢ for every $1 of revenue in the third quarter of 2000 before paying overhead costs. The more it sold, the more it lost. The company blamed losses on delivery costs.

3. MORTGAGE.COM

During the low interest rates, the site was doing great refinancing people's houses. When interest rates increased, the business dried up, unable to get original loans. It closed its doors on October 31, 2000, laying off 513 workers.

4. RAZORFISH INC.

The New York–based information technology adviser saw a slight drop in profits, then sent out a mass e-mail that when asked to unsubscribe, re-e-mailed the recipients, eventually filling their mailboxes with each unsubscribe request. The company also announced a net loss of $24.9 million in its first quarter, and the loss of its subsidiary in Helsinki, dropping its stock even further.

The company's shares, which closed as high as $55 in February 2000, have closed under $1 for more than 60 consecutive sessions.

5. ETOYS.COM

Founded in 1997, eToys grew quickly, increasing its employee base from 13 to 235 people in 1998. In the end, eToys laid off about 700 workers, or 70% of its workforce, in January 2001. The company's European wing, which ranked as the top toy e-tailer in its market, also shut down, citing a lack of financial support from its parent company.

Early in February, eToys announced that it would file for bankruptcy, end operations, and lay off remaining employees. From $84.35 in October 1999, eToys closed at $13.06.

6. AMAZON.COM

Amazon.com chief executive officer Jeff Bezos reportedly said that he regrets the investments made in Living.com and Pets.com. According to Living.com's initial public offering (IPO) filing, which was later withdrawn, the site lost $46.5 million.

7. STARBUCKS

Coffee king Starbucks lost nearly $60 million on Internet-related investments like the now-closed Living.com and Kozmo. Kozmo.com was a one-hour delivery service based on late-night runs. CEO Gerry Burdo announced that the company was shutting down in all of its nine cities—putting 1,100 people out of work, about 475 of them in the home office in Manhattan.

8. KNIGHT-RIDDER

The media giant lost $168 million on investments such as webvan.com, GoTo.com, and Infospace. Webvan.com, the online grocer, ended up selling millions in assets, including kitchen equipment, delivery vans, and warehouses.

9. PRICELINE.COM

The name-your-price site suffered a net loss of $105 million, or 62¢ a share, compared with losses of $921.4 million a year earlier. At one time Priceline reported an operating loss of $25 million, or 15¢ per share, more than doubling the loss analysts were anticipating. Chief executive Daniel Schulman blamed the higher-than-expected loss on seasonal weaknesses, costs associated with the closing of its gasoline and grocery sales services, and negative news stories about customer satisfaction. Yet he expected, and saw, an upturn with the $73 million investment from Hong Kong's richest tycoon, Li Ka-shing.

10. **IDEALAB!**

Another bad idea was beauty site Eve.com, which burned through $28.7 million in barely over a year before closing in October 2000. Idealab! then closed some of its offices and canceled its proposed IPO. The dot-com incubator behind eToys, CarsDirect, and Petsmart.com moved some of its Silicon Valley functions to its Pasadena, California, headquarters.

Bibliography

Abel, Ernest L. *Ancient Views on the Origins of Life.* Cransburg, N.J.: Associated University Presses, 1973.

Agosta, William. *Thieves, Deceivers and Killers: Tale of Chemistry in Nature.* Princeton, N.J.: Princeton University Press, 2001.

"Alien Life on Earth." *Popular Mechanics.* December 1996.

Altman, Roberta. *Every Woman Handbook for Preventing Cancer.* New York: Pocket Books, 1996.

Armstrong, W. P. "Symbiosis in Cycads." *Pacific Horticulture.* 1 November 2001 <http://waynesword.palomar.edu/ww0803.htm#cyanobacteria>.

Baker, David. *Inventions from Outer Space: Everyday Uses for NASA Technology.* New York: Random House, 2000.

Bakker, Robert T. *The Dinosaur Heresies.* New York: William Morrow, 1975.

Barnard, Jeff. "Humongous Fungus." 6 August 2000. Associated Press. 1 November 2001 <http://abc.news.go.com/sections/science/DailyNews/fungus000806.html>.

"The Biggest Living Thing on Earth." American Society for Microbiology. 1 November 2001 <www.microbe.org/news/giant_fungus.asp>.

Blakey, Elizabeth. "Venture Capital's Biggest Losers." 26 April 2001. NewsFactor Network. 1 November 2001 <www.newsfactor.com/perl/story/9180.html>.

Boss, Alan. *Looking for Earths: The Race to Find New Solar Systems.* New York: John Wiley & Sons, 1998.

Brain, Marshall. "How Domain Name Servers Work." HowStuffWorks Inc. 1 November 2001 <www.Howstuffworks.com>.

Brecker, Kenneth, and Michael Feirtag. *Astronomy of the Ancients.* Cambridge, Mass.: The MIT Press, 1974.

Brody, Jane. *Jane Brody's Cold and Flu Fighter.* New York: WW Norton, 1995.

Caper, Jean. *Stop Aging Now!* New York: HarperCollins, 1995.

Clark, Oliver. *Never Catch a Cold Again.* Englewood Cliffs, N.J.: Prentice-Hall, 1979.

Cobb, Cathy, and Harold Goldwhite. *Creations of Fire.* New York: Plenum, 1995.

Daly, Martin, and Margo Wilson. *Sex, Evolution and Behavior.* North Scituate, Mass.: Duxbury, 1978.

Davenport, Penny. *The Natural Way with Colds and Flu.* Shaftesbury, Dorset, England: Element Book Ltd., 1995.

Dwyer, Douglas. "How Atomic Clocks Work." HowStuffWorks Inc. 1 November 2001 <www.Howstuffworks.com>.

The Economist (US) 21 December 1996, 111.

Felton, Bruce, and Mark Fowler. *Best, Worst, and Most Unusual.* New York: Thomas Y Crowell, 1975.

Freudenrich, Craig C. "How Air Traffic Control Works." HowStuffWorks Inc. 1 November 2001 <www.Howstuffworks.com>.

Fulder, Stephen. *An End to Aging?* New York: Destiny Books, 1983.

Greene, Mott T. *Geology in the Nineteenth Century.* Ithaca, N.Y.: Cornell University Press, 1982.

Gribbin, John. *Companion to the Cosmos.* Boston: Little, Brown and Company, 1996.

Harris, Marvin. *Cannibals and Kings.* New York: Random House, 1977.

Hathaway, Nancy. *Friendly Guide to the Universe.* New York: Viking, 1994.

Hodge, Andrew. *Alan Turing: The Enigma.* New York: Simon & Schuster, 1983.

Hopkins, Jerry. *Strange Foods.* Tokyo: Periphus Editions Ltd., 1999.

Hunter, Mark. *Fantastic Journeys: Five Great Quests of Modern Science.* New York: Walker and Company, 1980.

Kass-Simon, G., and Patricia Farnes. *Women of Science: Righting the Record.* Bloomington, Ind.: Indiana University Press, 1990.

Lane, Dr. I. William, and Linda Comac. *Sharks Don't Get Cancer.* Garden City Park, N.Y.: Avery Publication Group, 1992.

Lewis, Cherry. *The Dating Game: One Man's Search for the Age of the Earth.* Cambridge, U.K.: Cambridge University Press, 2000.

Lyon, William F. *Insects as Human Food.* Columbus, Ohio: Ohio State University Press, 1996.

Mahoney, Michael. "Idealab! Jumps Silicon Valley Ship." 8 March 2001. E-Commerce Times. 1 November 2001 <www.EcommerceTimes.com>.

McGayne, Sharon Bertsch. *Nobel Prize Women in Science.* Carol Stream, Ill.: Carol Publishing Group, 1993.

McSween, Harry Y. *Fanfare for Earth: The Origin of Our Planet and Life.* New York: St. Martin's Press, 1997.

Mitchell, Richard Scott. *Mineral Names: What Do They Mean?* New York: Van Nostrand Reinhold, 1979.

Nash, J. Robert. *The World's Greatest Eccentrics.* Piscataway, N.J.: New Century Publishers, 1982.

Oesper, Ralph E. *The Human Side of Scientists.* Cincinnati, Ohio: University Publications, 1975.

Ogilvie, Marilyn Bailey. *Women in Science: Antiquity through the Nineteenth Century.* Cambridge, Mass.: The MIT Press, 1993.

People's Medical Publishing House, comp. *The Chinese Way to a Long and Healthy Life.* New York: Hippocrene Books, 1984.

Pelton, Ross, and Lee Overholsen. *Alternatives in Cancer Treatment.* New York: Simon & Schuster, 1994.

"Phaeophyta: Life History and Ecology." The University of California Museum of Paleontology. 1 November 2001 <www.ucmp.berkeley.edu/chromista/browns/phaeolh.html>.

"Phenylpropanolamine FDA Warning." Boston: Phenylpropanolamine Legal Network, 1 November 2001 <www.phenylpropanolamine-ppa.com>.

Pickover, Clifford A. *Strange Brains and Genius: The Secret Lives of Eccentric Scientists and Madmen.* New York: HarperCollins, 1999.

Piller, Charles. "Power Grid Vulnerable to Hackers." *Los Angeles Times.* August 13, 2001.

Pyctor, Helena M., and Nancy Slack. *Creative Couples in the Sciences.* New Brunswick, N.J.: Rutgers University Press, 1996.

Raup, David M. *Extinction, Bad Genes or Bad Luck?* New York: W. W. Norton, 1991.

Rugoff, Milton, and Ann Guilfoyle, eds. *The Private Lives of Animals.* New York: Grosset and Dunlop, 1974.

Salzberg, Hugh W. *From Caveman to Chemist.* Washington, D.C.: American Chemical Society, 1991.

Secrets of the Alchemists. Alexandria, Va.: Smithsonian Institute Libraries, Time-Life Books, 1990.

Spalding, David A. E. *Dinosaur Hunters.* Rockling, Calif.: Prime Publishers, 1993.

Sparks, John. *Battle of the Sexes.* London: BBC Books, 1999.

"Species Profiles." 10 April 2001. National Agricultural Library of the U.S. Department of Agriculture. 1 November 2001 <www.invasivespecies.gov/profiles/main.shtml>.

Stanley, Steven M. *Extinction.* New York: Scientific American Library, 1987.

Stark, Lloyd R. *The Ultimate Cause and Preventive Cure for the Common Cold.* Henderson, Nev.: Mojave Publications, 1994.

Taubes, Gary. *Bad Science: The Short Life and Weird Times of Cold Fusion.* New York: Random House, 1993.

"Tomatoe History." 8 May 2000. Lynchburg Community Market. 1 November 2001 <www.lynchburgmarket.com/TomatoeFaire.htm>.

Turkington, Carol. *Natural Cures for the Common Cold.* Gig Harbor, Wash.: Harbor Press, 1999.

Whelan, Elizabeth. *The Complete Guide to Preventing Cancer.* Amherst, N.Y.: Prometheus Books, 1994.

Williams, Robyn. "Nanobes." 3 March 2000. *Radio National.* 1 November 2001 <www.abc.net.au/rn/science/ss/stories/s132235.htm>.

"World's Largest Living Organism Found." 6 August 2000. *Richmond Times Dispatch.* 1 November 2001 <www.vgspc.com/newsy/armillaria.htm>.

Index

Abishaq of Shunem, 260
Acritarch algae, 194
Adams, Jon C., 85
Advanced Research Projects Agency Network (ARPANET), 285
Afrovenator, 207
Aibo, 271
Air baths, 256
Air traffic control, 284
Airline reservation systems, 288
Airtran. *See* ValuJet Airlines
Albertus Magnus, 232
Albrecht, Andreas, 100
Alcock, John, 177
Aldrin, Edwin Eugene Jr., 155
Algol, 61
Al-Kindi, 230
Alsopp & Sons, 7
Altavista.com, 294
Alternating current, 26–27
AlterNIC, 291–92
Altman, Roberta, 247
Alvarez, Luis, 211–12
Alvarez, Walter, 211–12
Alzheimer, Dr. Alois, 58
Amadeus, 288
Amazon.com, 296
Amethyst, 155
Anaximander of Miletus, 129, 130
Anaximenes, 130
Anderson, Elizabeth Garrett, 43
Andrews, Roy Chapman, 203–4
Androniclus, 94
Antifog spray, 266
Antiparticle, 113
AOL, 281
Apollo, 267–68, 275
Apollo/Galileo reservation system, 288
Aqua vitae, 240
Aractica, 145
Archaea, 187
Archaeocyathids, 194–95
Archaeopteryx, 204
Archbishop Ussher, 138
Archelaus of Athens, 131
Ardipithecus ramidus kadabba, 217–18
Ardipithicus, 218
Argentinian ants, 169
Argentinosaurus, 207, 208
Aristotle, 117–18, 119, 131, 224, 230
Armacolite, 155
Armillary sphere, 94
Armstrong, Neal Alden, 155
Arp, Halton, 74–75, 104
ARPANET. *See* Advanced Research Projects Agency Network
Asbestos, 155
Asian, Dr. Ana, 262
Asteroid, 211–12
ASTP program, 275
Astrariium, 95
Astro Zombies, The, 18

Index

Atomic clock, 286–87
Atomic Energy Commission, 66, 69
Atomists, 131
Atrolabe, 95
Attrition.org, 291
Augusts D., 58
Australopithicus africanus, 215–16
Australopithicus robustus, 216
Australopithicus, 54, 215–16, 217
Avicenna, 224, 226
Axial precession, 150–51
Ayrton, Hertha Marks, 42
Ayrton, W.E., 42

Baby HAL, 269–70
Bacon, Francis, 6
Bacon, Roger, 233
Baeyer, Adolf von, 8
Bain-marie, 231
Bakker, Robert T., 205
Baldwin, E., 210
Baltica, 145
BankAmericard, 286
Banting, Sir Frederick Grant, 58–59
Barnum, Phineas T., 203
Barometric pressure, 191
Baryon, 114
Bassett, Charlie, 274
BAT chord keyboard, 266
Baumbardner, John, 149
Beast of Yucca Flats, The, 18
Becher, Johann Joachim, 118
Becquerel, Antoine-Henri, 40
Bee colonies, 177
Bees, killer, 169–70
Belladonna, 182
Beringer, Dr. Johann, 19
Berthelot, Pierre-Eugene-Marcelin, 44
Berthollet, Claude-Louis, 243
Best, Charles Herbert, 58–59
Bethe, Hans, 124
Bighorn Medicine Wheel, 93–94
BizBuyer.com, 294
Black holes, 103, 105, 135
Black skull, 216
Bohr, Aage, 121–22
Bohr, Niels, 121–22
Bonaparte, Jose, 207
Bondi, Hermann, 99
Bose, Satyendra Nath, 113

Boson, 113
Brachiopods, 195
Brahe, Tycho, 80–81, 82, 95
Brainhat, 271
Brand, Vince, 275
Branes, 101, 126
Branies, Jay A., 135
Bride of the Monster, 17
Brooks, R., 199
Brot, J.B., 7
Brown and black rat, 172
Brown, Bardnum, 203
Browsers, 281
Bruno, Giordano, 34
Bucket orchid, 237–38
Bullfrog, 174
Bulls' genitals, 260
Bunsen, Robert Wilhelm, 8, 12, 62, 156
Bunsenite, 156
Buran-Energia, 276
Burdo, Gerry, 296
Burnham, S.W., 89
Bush tucker grubs, 190
Butenandt, Adolf, 73
Butler, Paul, 86, 87, 92, 106, 108

Caisse, Rene, 253
Calaveras skull, 20
Calculus, 25
Calhoun, John C., 157
Caloric, 119
Calvin, Melvin, 63–64
CAM Brain Machine, 272
Cambrian era, 194–95
Campbell, William Wallace, 45
Cannon, Annie Jump, 41, 62
Caper, Jean, 262
Carcharidontosaurus, 209
Cardboard palm, 237
Carnotaurus, 206
Carradine, John, 18
CarsDirect, 297
Carson, Rachel, 249
Castor-oil plant, 181
Caterpillar fungus, 260
Cavendish, Henry, 11–12, 228
Cepheid Variable, 98–99
Cernan, Eugene, 266
CGI. *See* Common gateway interface

Index

Chadwick, James, 72–73
Challenger, 276
Chamberlain, Thomas, 139
Chapparal, 253
Charles I, 261
Charles X, 6
Charlier, Charles, 97
Charm, 115. *See also* Quark
Chernobyl, 250
Chlorine gas, 257
Chricton's ankylosaur, 205
Chromium, 262
Church bells, 192
Cicadas, 192
Cifelli, Dr. Richard, 208
City living, 250
Clark, Oliver, 256
Climactic buzz, 152
Clinton, Bill, 68
CMGI, 294
Cochran, William, 87
Coconut larvae, 189
Collins, Michael, 155
Columbia, 276
Columbite, 156
Combs, Jeffrey, 18
Common gateway interface (CGI), 281
Computer Decency Act, 290
Comstock, Anna Botsford, 52
Comstock, John Henry, 52
Comte de Buffon, 138, 146
Connery, Sean, 17–18
Conodonts, 195
Coogan, Jackie, 17
Cope, Edward Drinker, 27–28, 200–1, 203
Copernicus, Nicolaus, 80
Copper, 253
Cori, Carl, 37, 52
Cori, Gerty Radnitz, 37, 52
Coria, Rudolfo, 207
Cosmonauts, female, 275–76
Cottonwood tree, 192
Couper, Archibald Scott, 57
Cowbird, 171
Credit card verification, 286
Cretaceous era, 194, 196, 207, 208, 209, 211
Crichton, Michael, 205

Crick, Francis, 60, 134
Croll, James, 150–51
Cro-Magnon, 218–19
Cronkite, Walter, 33
Crylophosaurus, 206–7
Cupping, 258
Curie, Bronya, 40
Curie, Irene, 41, 52–53
Curie, Marie Skodowka, 36, 40–41, 52, 157
Curie, Pierre, 40–41, 52, 157
Currant, Andrew, 21
Cycads, 186–87

Daisies, 192
Dalton, Francis, 12, 121
Dalton, John, 227–28
Dark matter, 104–5
DART. *See* Dextrous Anthropomorphic Robotic Testbot
Darwin, Charles, 24, 57, 142
Davidson, George, 89
Davidson, Norman, 125
Davis, Joe, 9–10
Davy, Sir Humphrey, 36, 228
Dawson, Charles, 20–21
DDT, 249
De Beaumont, Elle, 142–43
De Dondi, Giovanni, 95
De Garis, Hugo, 270
De Laplace, Pierre-Simon, 227
De Leon, Juan Ponce, 259
De Maillet, Benoit, 133
Death Star. *See* Nemesis
Debentox, 21
Deccan Traps, 211
Dehmelt, Hans, 59
Deinodon, 203
Dekker, Albert, 16
Delambre, Andre, 17
Democritus, 117, 131
Desaga, Peter, 156
Descartes, Rene, 137
Devonian era, 195
Dextrous Anthropomorphic Robotic Testbot (DART), 271–72
Diffendaffer, Paul, 267
Diodorus Siculus, 132
Dioxin, 250–51
Dirac, Paul, 115–16

Direct current, 26–27
Discovery, 91
DNS. *See* Domain names server
Domain name server (DNS), 285
Domain names, 280
Dong Zhiming, 205
Dor, R. H., Jr., 210
Douchez, Dr., 257
Dowd, Brian, 271
Dr. Thorkel, 16
Dr. Cyclops, 16
Dr. Jekyll and Mr. Hyde, 16
Dr. Marco, 18
Dr. No, 17–18
Dr. Strangelove, 18
Drafts, 257–58
Dragonflies, 190
Drudge, Matt, 292–93
Duck World, 290
Dumas, Jean-Baptiste-Andre, 6, 243
Dung beetles, 189
Duquennoy, Antoine, 86–87

Ediacara fauan, 194
Edison, Thomas Alva, 26–27, 45
Einhorn, Alfred, 262
Einstein's Rings, 105
Ekpyrotic universe, 136
Electron, 113, 114, 120
Electronic use, as cause of cancer, 249
Electronic water filter, 268
Electrostatis bonding theory, 243
Elizade, Manual Jr., 21
E-mail, 279, 280
Emanuel, Kerry A., 213
Embryonic diapause, 175–76
Empedocles of Acragus, 130–31
Encelades, 88
Endolithic communities, 184–85
English ivy, 167–68
Epsilon Eridani, 108
Epsilon Reticulum, 109
ER 1470, 217
Erus, Justis, 217
Escherida coli, DNA inserted into, 10
Essiac, 253
Eta Carinae, 102–3
Etoposide, 253

eToys.com, 295–96
Eve.com, 297
Exoplanets, 106–9
Extraterrestial dust, 152

Facial tissues, 257
Faraday, Michael, 36, 48, 228
Farquhar, Marilyn, 54
Federal Open Market Committee (FOMC), 288
Federal Reserve, 287–88
Fermi, Enrico, 113
Fermi-Dirac statistics, 122
Fermion, 113, 116, 122
Feynman, Richard, 14, 124–25
Fifty one Pegasus, 108
First type theory, 243
Fischer, Emil Hermann, 37
Flavors, 115
Fluegge, Siegfried, 123
Fluffy Bunny, 291
Fly, The, 17
FOMC. *See* Federal Open Market Committee
Fong, Peter, 5
Foraminiferans, 196
Fountain of Youth, 259
Fourcroy, Antoine-Francois, 242–43
Foxglove, 180–81
Frankenstein, Baron, 16
Franklin, Rosalind, 59–60
Frederick II, 81, 95
Freedom 7, 274
Free-lunch universe, 135
Freeman, Derek, 31–32
Freud, Sigmund, 28–30
Friedmann, Alexander, 98
Frisch, Otto, 123
Fuchs, Klaus, 125
Fuller, Buckminster, 50
Furniture.com, 294

Gagarin, Yuri, 273, 274
Galen, 226, 236, 240
Galileo Galilei, 24–25; 70, 80, 84, 96
Galle, Johann Gottfried, 85
Gallimimus bullaturs, 209
Galvani, Luigi, 241
Gaposchkin, Segei, 53

Index

Gardiner, Brian, 21
Gardner, Marshall B., 143–44
Garnet, 156
Gary, Fred, 68
Gatewood, George, 86
Gay-Lussac, Joseph-Louis, 7
Gelm, Andre, 4
Gemini flights, 266, 274
Geneva Observatory, 108
George III, 85
Georget, D.M.R., 3
Gerbils, 176–177
Ghost Shirt Society, 291
Gibbs, Josiah Willard, 45
Giganotosaurus, 207
Ginger, 261
Ginseng, 261
Glenn, John, 273
Gliese 876, 107
Gold, Thomas, 99
Golden Gate Bridge, 268
Goodricke, John, 61
Gordon, Stuart, 18
Grahan, Thomas, 7
Graphical user interface, 278
Grass carp, 170
Grasshoppers, 189
Graviton, 115
Great Barrier Reef, 175
Great Pyramid at Giza, 92
Great Wall, 105
Greek fire, 239–40
Grilling, as cause of cancer, 247
Grissom, Gus, 273
Gross, Al, 267
Gutenberg, Beno, 50
Guth, Alan, 99–100, 135

Haber, Fritz, 9
Hackers, 290
Hacking for Girliez (HFG), 292
Hahn, Otto, 71–72
Haile-Selassie, Yohannes, 217–18
Hair dye, 247–48
HAL 9000, 269–70
Hale reflector, 90–91
Hale, George Ellery, 89–91
Halley, Edmund, 26, 82, 137–38
Hallucinogenic mushrooms, 259

Hammer, Dr. William, 206–7
Hard drive, 278
Hartmann, Johannes, 226
Harvey, Sir William, 11, 261
Hashhashins, 224
Hatshepsut, 230
Hatzes, Artie, 87
Haughton, Samuel, 139
Hawking, Stephen, 64, 135
Hayden, Ferdinand Vandiveer, 202–3
Haynes, Suzanne G., 248
Hays, James, 151
Hazen, Robert M., 135
HD 168443, 106
HD 209458, 106–7
HD 210277, 107
HD 83443, 108
Heaviside, Oliver, 12–13
Heavy oxygen, 151
Hedison, David, 17
Hemlock, 180
Henbane, 183
Heraclitus, 130
Herschel, Caroline Lucretia, 39
Herschel, Sir John, 20, 39, 102–3
Herschel, Sir William, 39, 82–83, 85, 88, 102
Hertz, Rudolf, 50
Hess, Harry Hammond, 145
HFG. *See* Hacking for Girliez
Hierarchical model, 97
High-capacity removable storage, 279
Himalayas, 152–53
Hinton, Martin A.C., 21
Hipparchus, 79
Hodgkin, Dorothy Crowfoot, 63
Hoffman, A.W., 7
Hoffman, Erich, 45
Holmes, Arthur, 139–40, 148
Homo erectus, 217
Homo habilis, 216–17
Homo neanderthalis, 218–19
Homo rudolfensis, 216–17
Homo sapiens, 218, 219
Homosexuality, 248
Homunculus, 102–3
Honey mushroom, 187
Honeybees, 237
Honeypot ant, 188–89

Index

Honeysuckle, 163
Hooke, Robert, 25–26
Horse, 172–74
Houtermans, Fiesel, 140
Hoyle, Fred, 99, 34
Hubble Space Telescope, 91, 276
Hubble, Edwin Powell, 98–99
Huber, Claudia, 133
Huggins, William, 51
Hurricanes, 213
Hutton, James, 141, 146–47
Huygens, Christiaan, 84–85
Hydrilla, 166–67
Hydrogen bomb, 122–23
Hylaesosaurus, 200
Hyperon, 114
Hypertext markup language, 280
Hyperthermia, 253–54

Ice age theories, 150–54
Idealab!, 297
Ig Nobel Prize, 3–5
Iguanadon, 200
Imrie, John, 151
Inflation model, 99–100
Infogene, 10
Interacting galaxies, 104
Internet service providers (ISP), 281
Internet, 279
Invasive animals, 164–74
Invasive plants, 161–68
Island of Dr. Moreau, The, 16
Island of Lost Souls, The, 16
ISP. *See* Internet service providers
Iten, Oswald, 21

Jabir ibn Hayyan, 223–24
Jackson, Andrew, 157
Jacobi, Abraham, 51
Jacobi, Mary Putnam, 51
Jai Singh II, 95–96
Jaipur Observatory (Yantra), 95–96
Jansky, Karl, 90
Jex-Blake, Sophia, 43
Jimsonweed, 180
Jimthomsonite, 156
Johanson, Donald, 32–33
Johnson, Laurie, 267
Joliot-Curie, Frederic, 52–53

Joly, John, 138, 144, 147
Jonson, Tor, 18
Josemithite, 156
Jujube berries, 261
Jung, Carl Justav, 28–30
Jurassic era, 207, 208
Jurassic Park: The Lost World, 290

K/T boundary, 211–12
Kangaroo, 175–76
Kashpureff, Eugene, 291–92
Keck, W. M., observatory, 91–92
Kelp, 186
Kenyanthropus platyops, 217
Kepler, Johannes, 61, 81–82, 88
Keratakis, 231
Khoury, Justin, 135
Kibble, T.W.B., 100–101
Kimberly Clark, 257
King David, 260
King, Helen Dean, 41
Kinway, Richard, 141
Kirschvinik, Joseph L., 153
Knight Ridder, 296
Ko Hung, 231
Komarov, Vladimir Mikhailovich, 273–74
Kozmo.com, 296
Kraepelin, Emil, 58
Kriegsman Furs and Outerwear, 291
K-strategy, 214
Kubrick, Stanley, 18
Kubrick, Stanley, 31
Kudzu, 162

Lake Vostok, 184
Lake, Matt, 289
Lalande, Joseph-Jerome de, 86
Lambornella clarki, 238
Landsat, 265
Langeland, Kenneth, 167
Lascu, Christian, 185
Laughton, Charles, 16
Laurent, Auguste, 243
Lavoisier, Antoine-Laurent, 26, 34–36, 56, 119, 227, 242
Le Bel, Joseph-Achille, 57–58
Lead paint study in Baltimore, 68–69
Leafy spurge, 164–65

Leakey, Louis, 32–33, 53–54, 216
Leakey, Mary Douglas Nicol, 32–33, 53–54
Leakey, Richard, 32–33, 217
Leibniz, Gottfried Wilhelm, 24, 25, 55
Lek, 176
Lepton, 113, 114, 115, 136
Leucippus, 117
LeVerrier, Urbain, 85
Lewis, Jerry, 17
Lewis, W. Lee, 9
Lewisite, 9
Li Ka-shing, 297
Libavius, 240–41
Lick Observatory, 89
Lick, James, 89
Liebig, Justus von, 8
Lina, Don, 275
Linde, Andrei, 99–100
Lippershey, Hans, 96
Lister, Joseph, 44
Lithographiae Wirceburgensis, 19
Living.com, 296
Locke, Richard Adams, 20
Lonsdale, Kathleen Yardley, 53
Lonsdale, Thomas, 53
Lovecraft, H.P., 18
Lowell, Percival, 83
LSD tests, 67
Lucretius Carus, 132
Lucy, 32–33, 216
Lugosi, Bela, 17
Lyell, Sir Charles, 142

MacDonald, Gordon J., 152
Mach, Ernst, 49
Macleon, John James Rickard, 58–59
Magellanic Clouds, 104
Magnetic fluid injection, 268
Mainframes, 277
Makino, Takeshi, 5
Makovicky, Dr. Peter J., 209
Mallee fowl, 175
Manhattan Project, 14, 66, 121–25
Mantell, Dr. Gideon, 199–200
Marat, Jean-Paul, 26, 36
Marconi, Guglielmo, 50
Marcy, Geoffrey, 86, 87, 92, 106, 107, 108

Maria the Jewess, 231
Markoff, John, 292
Marooned, 274
Marsh, Othniel Charles, 27–28, 202–3
Mawson, Sir Douglas, 194
May apple, 253
Mayor, Michel, 86–87, 108
McBride, William, 21
McClintock, Barbara, 75
McLeod, Dewey M., 211
Mead, Margaret, 31–32
Megalosaurus, 199
Meitner, Lise, 71–72
Mendeleev, Dmitri, 44, 114
Mengele, Dr. Josef, 65–66
Mercury flights, 274
Mercury poisoning, 231
Mercury, 107
Mesa of Lost Women, 17
Meson, 114
Mesozoic era, 214
Meteorite theory of extinction, 205
Methane, 212
Metzinger, Polly, 22
Meyer, Julius Lothar, 119
Microetching, 267
Microvenus, 10
Midpixel Digital Mind Modeling Project, 272
Mikels, T.V., 18
Milanivitch, Milutin, 151
Mile-A-Minute, 164
Miller, Stanley, 134–35
Mimas, 88
Minnows, fathead, 235–36
Mir, 275–76
Mirkwood, Galadriel, 22
Missing link, 22–23
Mistletoe, 253
Mitchell, Maria, 42
Mitchell, William, 42
Mitnick, Kevin, 289, 292, 293
Molly, Amazon, 176
Monopole universe, 100
Moorbath, Stephen, 149
Moore, John, 68
Morris, Robert Tappan, 289–90
Mortgage.com, 295
Mosquitoes, 238

Index

MotherNature.com, 294
Mount Wilson Observatory, 89–90
Mouse, 278
Mouthwash, 248
Moville Cave, 185
MSN, 281
M-theory, 101
Mud slide, 154
Muller, Richard, A., 152
Mullis, Kary, 15
Multimedia, 279
Muon, 113, 114
Murex brandaris, 235
Murphy, Eddie, 17
Mushroom, *Paxillus Involutus*, 179–80

Nanobes, 186
Napoleon, 242
Nastulus, 95
National Aviation and Space Administration, 265–68
Nautiloids, 195
Nectar, 237
Nemesis (Death Star), 103
Nena, 145
Neptune, 103
Network Solutions (InterNIC), 291–92
Neutrino, 113, 114, 115
Neutron, 115
New York Sun, 20
New York Times, The, 292
Newell, Kenneth, 4
Newton, Sir Isaac, 24–26, 47, 55, 61, 82, 84, 88, 138
Nicolle, Charles J.H., 63
Nolan, James F., 4
Nothronycus, 207–8
Noyes, Robert, 87
Nuclear magnetic resonance, 265
Nuclear power plants, 249–50
Nutty Professor, The, 17

Okamura, Chonosuke, 3
Oleander, 181
Olsen, Paul, 153–54
Olshevsky, George, 206
Oparin, Alexander, 134
Open universe, 98
Operation DNS Storm, 292

Oppenheimer, J. Robert, 30–31, 73, 74, 122–23, 124
Optical illusions, 191
Ordipithecus ramidus, 215
Ordovician era, 195
Oriental medicines, 261
Ornithominid, 209
Osborn, Henry Fairchild, 204
Out of Africa model, 218
Owen, Sir Richard, 200, 206
Ozone, 253

Pagels, Dr. Heinz, 38
Pakjad, Bijan, 4
Palade, George, 54
Pancake universe, 98
Panspermia, 134
Paracelsus (Bombast von Hoehnheim), 118, 224–25, 240
Paralitan stromeri, 208–09
Paranthropus, 216
Parker, A.C., 3
Parkinson, John, 236
Parr, Thomas, 261
Parson, William (Earl of Rosse), 89
Pasteur, Louis, 244
Patterson, Clair, 140, 148–49
Paul, Gregory S. 213–14
Pauli, Wolfgang, 59, 114
Payan, Anselm, 7
Payne-Gaposchkin, Cecilia, 53
Peabody, George, 202
Pederson, Paul, 268
Peierls, Rudolf, 123
Pelorosaurus, 200
Penrose, Roger, 135
Periodic Table, 119
Perl, Martin, 115
Permian era, 195
Personal computers, 277–78
Pets.com, 295
Petsmart.com, 297
Phenylpropanolamine, 256
Philosopher's stone, 232
Philosophical essences, 118
Phogiston, 118, 119
Pickering, Edward, 41
Piller, Charles, 284
Piltdown man, 20–21

Index

Plague, bubonic, 66–67
Planck time, 99
Planet X, 103
Plant oil, 191
Plateosaurus, 204
Pleistocene era, 193
Pleistocene Megafauna kill, 193
Plot, Robert, 199
Plum pudding model of atoms, 120
Po Lin Buddha, 268
Polaris submarine, 122
Polychlorinated biphenyls, 248
Positron, 115–16.
Power grid, 284
Precambrian era, 194
Priceline.com, 296–97
Priestley, Joseph, 55–56, 70
Procaine, 261–62
Proconsul, 54
Protarchaeopteryx robusta, 208
Proton, 114
Proust, Joseph-Louis, 243
Prout, William, 119
Prozac, 5
Psynet model, 270
Ptolemy, 79, 94
Purple loosestrife, 165–66

Quaker Oats Company, 69
Quark, 114, 115, 116, 136
Queen Anne's Lace, 167
Queloz, Didier, 86–87, 108

Racial senescence, 211
Radiation blockers, 267
Raging Bull, 294
Raymo, Maureen, 152–53
Razorfish Inc., 295
Re-Animator, 18
Reber, Grote, 90
Reefs decimated, 195
Regnosaurus, 200
Relaxation, 258
Rial, Jose, 151
Rodinia, 145, 153
Rogers, John, 145
Roosevelt, Franklin D., 123
R-strategy, 214
Rudolph II, 81, 95

Runaway subduction, 149
Rutherford, Ernest, 147–48

Sabin, Albert, 46
Sabre reservation system, 288
Saenger, Eugene, 66
Salamanders, American, 177
Salgado, Leonardo, 207
Salk, Jonas, 46
Saltcedar, 163–64
Saltpeter, 233
Sands, John P. Jr., 4
Sargasso Sea, 186
Saturn, 88
Sauropseidon, 208
Savitskaya, cosmonaut, 276
Schaeffer, Julius, 179–80
Schaudinn, Fritz Richard, 45
S-Check, 5
Scheele, Carl Wilhelm, 55–56
Scheele, Karl Wilhelm, 36
Schulman, Daniel, 296–97
Schwiner, Julian S., 125
Sciama, Denis, 99
Science Club at Fernald School, 69
Sea monsters, 196
Sea prawns, 259–60
Search engines, 281
Second type theory, 243
Secure socket layers (SSL), 281
See, Elliot, 274
Self-healing computers, 266
Sellers, Peter, 18
Sereno, Paul, 207
Shackleton, Nicolas, 151
Shapley, Harlow, 98
Shark cartilage, 253
Shepard, Alan, 274
Shimimura, Tsutomu, 292
Shoemaker, Eugene, 37, 212
Shrimp, 185
Shunamitism, 260–61
Shyness, 257
Sidoll, Dr. Mara, 5
Silkworm caterpillars, 190
Singularity theory, 135
Size, 213–14
Skodoskite, 157
Skylab, 274–75

Slipher, Vesto Melvin, 99
Slutsky, Robert, 22
Smith, A.C., 3
Smith, Joseph V., 156
Smith, Joshua, 208
Smithson, James, 157
Smithsonite, 157
Snake Pit vent, 185
Sneaker gel, 267
Snowball Earth, 153
Social Security numbers, 287
Soddy, Frederick, 147–48
Sokal, Allan, 22
Sony Inc., 271
Soybeans, 253
Soyuz, 273–74, 275
Spider, Redback, 177–78
Spiders, 189, 192
Spinoza, Baruch, 85
Sprigg, R.C., 193
SSL. *See* Secure socket layers
Stadonitz, Friederich August Kekule von, 57
Standard model of cosmology, 99
Starbrain Project, 272
Starbrain, 270
Starbucks, 296
Stark, Lloyd R., 257–58
Starlab, 270
Starling, 172
Starling, Ernest Henry, 45–46
Statue of Liberty, 268
Steady-state model of the universe, 99
Steinhardt, Paul, 100, 136
Steinmetz, Charles, 62
Sternberg, C.H., 203
Stevems, Nettie Maria, 42
Stick insects, 236
Stillwell, Thomas J., 4
Stock markets, 283
Stonehenge, 93
Stringer, Christopher, 218–19
Stromer, Ernst, 208–9
Strutt, R.J., 148
Suess, Eduard, 143
Suk, Joseph, 99
Sulfur pearl of Namibia, 187
Sulfuric acid, 213
Superstrings, 100–101, 136

Swallows, 192
Sweet, Dr. William, 67–68
Syquest disk, 279
Szilard, Leo, 122

Tamarisk, 163–64
Tasaday, 21
Tau Bootes, 107
Tau particle, 113, 115
Teller, Edward, 18, 30–31, 63, 73, 123–24
Termites, 190
Terra computer model, 149
Tesla, Nikola, 13, 50
Thales of Miletus, 129–30, 223
Thalidomide, 21
Theophrastus, 131
Therapsids, 195
Thomas Aquinas, 232
Thompson, James B., Jr., 156
Thompson, William (Lord Kelvin), 48–49, 138–39, 147, 148
Thomson, Sir Joseph John, 120
Three Mile Island, 250
Titov, Gherman, 273
Tobacco, 236
Tombaugh, Clyde, 83, 85
Tomonaga, Shinichiro, 125
Tortoise shell, 261
Tower of the Winds, 94
Trachodon, 203
Tranquillityite, 157
Tria Prima, 118
Triassic era, 196, 204
Triceratops, 203
Trilobite, 194–95
Triquentrum, 94
Tryon, Edward, 135
Tulieres, 6
Turing Test, 269–70
Turing, Alan, 14–15
Turk chess automaton, 269
Turkingtoin, Carol, 258
Tuskeegee experiments, 68
Tyrannosaurus, 203, 207

U particle, 115
Ulysses spacecraft, 276
United Loan Gunmen, 292–93
Universal gravitation, 26

Upsilon Andromidae, 108
Uraniborg, 81, 95
Uranus, 103
Urban IV, 233
Urban VIII, Pope, 24–25
Uwins, Dr. Philippa, 186

ValuJet Airlines, 292
Van Helmont, Johan Baptista, 226
Van't Hoff, Jacobus, 57–58
Vannier, Dr. Michael, 265
Vavilov, Nikolay, 72
Venter, J. Craig, 187
Visa, 286
Vitruvius, 94
Vogt, Peter R., 211
Volcanism, 211
Volta, Alessandro Giuseppe Antonio Anastasio, 48, 242
Von Braun, Werner, 274
Von Hohenheim, Bombast. *See* Paracelsus
Von Huene, Friederich, 204

Wachtershauser, Gunter, 133, 136
Walcott, Charles, 147
Wallace, Alfred Russel, 24, 56
Wasps, ichneumon, 171
Water bugs, 189
Water chestnut, 161
Watson, James, 60, 134
Watt, James, 48
Webmind AI Engine, 270
Weevils, 237
Wegener, Alfred, 144
Weinberg, Steven, 99
Wells, H.G., 16
Werner, Abraham Gottlob, 142
West, Herbert, 18
Westinghouse, George, 27
Wetherell, David, 294
Wheatgrass, 253

Whelm, Elizabeth, 250
Whiston, William, 138
White snakeroot, 182–83
White, Tim, 33
Whitmore, Frank, 9
Whitney, J.D., 20
Wickramasinghe, Chandra, 134
Wilkins, Maurice, 59–60
Williams, Paul Jr., 4
Willis, Katherine, 152
Williston, Samuel Wendell, 203
Wilson, Jim, 187
Winged moon creatures, 20
Winograd, Isaac, 153
Wireless Internet, 282
Wiseman, Joseph, 17–18
Wolfe, Doug, 207
Wood, Ed, 17, 18
Wood, Robert, 8
Woodward, Arthur Smith, 21
Wool bleaching, 239
Word processing, 278
Worldspan, 288
Worm, The, 289–90
Wormholes, 105
Wormwood, 236
Wurtz, Charles-Adolphe, 57
WYSIWYG editors, 281

Xenophanes, 130

Yagyu, T., 5
Yahoo!, 281
Yantra. *See* Jaipur Observatory
Yew, 181–82
Youngberry, Charles, 267

Zebra mussels, 170–71
Zel'ddovich, Yakov, 99
Zinjanthropus, 54
Zosimus of Panopolis, 231–32
Zwicky, Fritz, 13, 104

About the Authors

Susan Conner is a senior editor for *Signs of the Times* magazine. She has written more than 500 articles on science-related topics and has won numerous awards for feature writing.

Linda Kitchen is a writer and editor also for *Signs of the Times*. She has degrees in biology and journalism and is training to become a veterinarian.

Both live in Cincinnati, Ohio.